D0913117

GOTTLOB FREGE
PHILOSOPHICAL AND MATHEMATICAL
CORRESPONDENCE

Uniform with this volume:

GOTTLOB FREGE
Posthumous Writings

GOTTLOB FREGE

Philosophical and Mathematical Correspondence

Edited by
GOTTFRIED GABRIEL
HANS HERMES
FRIEDRICH KAMBARTEL
CHRISTIAN THIEL
ALBERT VERAART

Abridged from the German edition by
BRIAN MCGUINNESS
and translated by
HANS KAAL

The University of Chicago Press

The University of Chicago Press, Chicago 60637

Published 1980
Printed in Great Britain by Spottiswoode Ballantyne Ltd.

Library of Congress Cataloging in Publication Data

Frege, Gottlob, 1848–1925.
    Philosophical and mathematical correspondence.
    Translation of selections from v. 2 of
Nachgelassene Schriften und Wissenschaftlicher
Briefwechsel.
    Includes bibliographical references and index.
    1. Mathematicians—Germany—Correspondence.
    2. Frege, Gottlob, 1848–1925.   I. Gabriel,
Gottfried, 1943–   II. Title.
QA3.F74213   1980   510   79-23199
ISBN 0-226-26197-2

Originally published in German under the title *Nachgelassene Schriften und wissenschaftlicher Briefwechsel*, vol. 2, by Felix Meiner (Hamburg, 1976).

TABLE OF CONTENTS

Correspondence

LIST OF ABBREVIATIONS USED

Numbers in angle brackets refer to the List of Works of Frege cited in this volume.

BS	= *Begriffsschrift* (*Conceptual Notation*) ⟨1⟩
BuG	= 'Über Begriff und Gegenstand' ('On Concept and Object') ⟨10⟩
E.T.	= English Translation
FB	= *Funktion und Begriff* ('Function and Concept') ⟨7⟩
GGA I	= *Grundgesetze der Arithmetik* I ⟨11⟩
II	= *Grundgesetze der Arithmetik* II ⟨17⟩
GLA	= *Die Grundlagen der Arithmetik* (*The Foundations of Arithmetic*) ⟨4⟩
GLG I	= 'Über die Grundlagen der Geometrie' ('On the Foundations of Geometry') ⟨18⟩

II	= ” ” ” ” ” ⟨19⟩
III 1	= ” ” ” ” ” ⟨21⟩
III 2	= ” ” ” ” ” ⟨22⟩
III 3	= ” ” ” ” ” ⟨23⟩
PW	= *Posthumous Writings* ⟨32⟩
SB	= 'Über Sinn und Bedeutung' ('On Sense and Reference') ⟨9⟩

HISTORY OF FREGE'S SURVIVING
CORRESPONDENCE*

On 27 July 1919 Gottlob Frege wrote to the chemist and historian of science Ludwig Darmstaedter:[1] 'Concerning the wish you expressed earlier that I leave the letters addressed to me to your autograph collection, I find it to be fully justified. I have started to sort those letters and to check which of them appear suitable for this purpose.' Frege did not, however, live to comply with Darmstaedter's wish. When Frege died on 26 July 1925 in Bad Kleinen, he left the letters from his most important scientific correspondents to his adopted son Alfred Frege along with his scientific manuscripts.

Frege's intention of making a suitable selection of letters, as evidenced by the above-mentioned letter to Darmstaedter, is also borne out by a provision in his will:[2] a disposition dating from the year 1919, the year he composed his letter to Darmstaedter, provides that only that part of the correspondence addressed to him which he himself had selected was to be handed over to Darmstaedter's autograph collection. Frege's disposition provided further that these letters could be published only thirty years after the death of each correspondent. With this provision Frege evidently meant to take care of a doubt which he had already expressed to Darmstaedter, namely whether he was not committing a 'breach of confidence' against his correspondents by leaving the letters to a public library.

On 26 November 1925 Alfred Frege then handed over to Darmstaedter's autograph collection, which was being continued within the framework of what was then the Prussian National Library, what was according to Scholz and Bachmann 'a not very extensive collection of letters to Frege' which

---

* Translated from the Editor's Preface to the German edition.

1 Lothar Kreiser's contribution, 'Zur Geschichte des wissenschaftlichen Nachlasses Gottlob Freges', *Ruch Filozoficznej* 33 (1974) No. 1, pp. 42–7, appeared too late to be considered in the following account. A comprehensive account of the scientific remains, which makes full use of all prior research and contains further information on its history, is to be found in Albert Veraart, 'Geschichte des wissenschaftlichen Nachlasses Gottlob Freges und seiner Edition', in Matthias Schirn, ed., *Studies on Frege*, vol. I, *Logic and Philosophy of Mathematics* (Stuttgart, 1976).

2 On this point cf. L. Kreiser's remarks on Frege's remains and biography in his review of *Nachgelassene Schriften*, vol. 1, in *Deutsche Zeitschrift für Philosophie* 21 (1973), pp. 521–4, esp. p. 521.

included as 'the centrepiece of the collection . . . ten letters by Bertrand Russell to Frege'.[3]

According to a communication from the German National Library dated 15.11.1960, the acquisition records of the Darmstaedter Collection show the following entry under No. 1925/104 and the date 27.11.1925: 'Letters and postcards from the remains of Prof. Frege, 20th cent. (121 items). From Frege, student, Bad Kleinen.' Unfortunately the Frege bundle did not remain together but, according to a communication from the German National Library dated 15.11.1960, 'was dissolved and distributed under the names of the individual correspondents, as was always done in the case of smaller collections.' A communication from the Manuscript Division of the National Library, Prussia's Cultural Heritage, dated 20.9.1974, about the number of items (121) states that, judging by the experience they have had, 'Darmstaedter [had] counted "items", that is, individual sheets, perhaps also envelopes, and certainly printed enclosures. [But] these data are no longer verifiable.' According to a communication from the National Library dated 16.10.1974, the remaining material was counted once more independently of the recorded figures: 'The individual sheets together with envelopes, visiting-cards, cuttings, and printed matter amount altogether to 113 "items". If we now also add the three letters that are lost [from Hilbert, 29.12.1899, Jourdain, 18.4.1914, and Hönigswald, 5.2.1925], we come close to "121 items".' It may therefore be assumed that all of the remaining correspondence handed over at the time by A. Frege to what is now called the National Library could be taken into account in preparing this edition.

The letters preserved in the Darmstaedter Collection were found by Heinrich Scholz in 1936, as part of his efforts to publish Frege's scientific remains. After he had found the letters *to* Frege, his main task was to locate the letters *from* Frege. At the International Congress of Scientific Philosophy in Paris in 1935, Scholz called on everybody to assist him in his search, and he repeated this call in the report composed by him and Friedrich Bachmann on 'The Scientific Remains of Gottlob Frege'.[4] As a result of his efforts Scholz succeeded in obtaining among other things the originals of Frege's letters to Russell. However, Scholz's efforts were not always crowned with success. Thus Wittgenstein (in a letter dated 9.4.1936) declined to put Frege's letters at his disposal because of their 'purely personal, not scientific content'. Couturat's widow told him that she had kept the remains of her husband's correspondence only for a while. Scholz seems to have obtained from Alfred Frege, together with a letter dated

---

3 Heinrich Scholz and Friedrich Bachmann, 'Der wissenschaftliche Nachlass von Gottlob Frege', *Actes du Congrès International de Philosophie Scientifique, Paris 1935*, vol. VIII, *Histoire de la Logique et de la Philosophie Scientifique* (Paris 1936), pp. 24–30.
4 Loc. cit.

16.1.1937, those letters *to* Frege that had not gone into the Darmstaedter Collection: 'Among the correspondence being sent to you by the same mail there are among other things a number of letters from Prof. Haussner, Jena.' A. Frege asks that Haussner's letters be returned to him because they are predominantly political and personal in character. A. Frege places no such condition on the other letters being sent along with them.[5]

It may therefore be assumed (1) that the other letters were also of scientific interest and (2) that Scholz had A. Frege's permission to keep them in his archive. The letters sent probably included all those letters *to* Frege that were marked '*Sch. Arch.*' in the notes prepared by Scholz and his collaborators, including *inter alia* Wittgenstein's letters to Frege. The notes on Frege's correspondence include three lists, under the title 'Frege's Scientific Correspondence', which will be called 'Scholz's lists' (*SchL* 1–3). Of these lists *SchL* 1 represents the latest state of Scholz's and his collaborators' efforts.

The originals of the letters collected by Scholz were burned when the city of Münster, including the University Library, was bombed on 25.3.1945.[6] Unfortunately Scholz had not made any photocopies of the originals which could have been kept at a different location. However, Scholz had made some typewritten copies of most of the letters for the edition he was planning. These transcripts, Scholz's lists, and the originals contained in the Darmstaedter collection have survived the war, and so have transcripts prepared by Frege's correspondents themselves (e.g. Dingler) or copies sent to them by Scholz at their request (e.g. Russell).

When we began the current edition we were also able to go back to these materials and sources. For the resumption of the efforts to publish Frege's scientific correspondence it was essential that the 'meta-correspondence' Scholz had been carrying on at the time on this subject had been preserved. The initiatives Scholz had taken but not carried out could thus be carried further, and part of what was lost could thus be reconstructed. Our research

---

5 In Alfred Frege's letter to Heinrich Scholz dated 16.1.1937 Alfred Frege mentions that the non-mathematical part of the remains of his adoptive father contains also among other things a 'correspondence with Prof. Schäfer, Berlin'. The possibility cannot be excluded that the Schäfer referred to is the historian Dietrich Schäfer (1845–1921) whose political views were quite close to Frege's. However, inquiries concerning Schäfer's remains have produced only negative results. Thus the State Archive, Bremen (in a letter dated 25.2.1975), and the Academy Archive of the Academy of Sciences of the German Democratic Republic (in a letter dated 19.3.1975) wrote that that part of Schäfer's remains which was under their care did not contain any letters from Gottlob Frege.

6 Cf. H. Hermes, F. Kambartel and F. Kaulbach, 'Geschichte des Frege-Nachlasses und Grundsätze für seine Edition', in *Nachgelassene Schriften*, vol. I, p. XXXVI.

was often quite time-consuming in its details, but it substantially enriched the original material for the edition.

In the course of our investigations the materials that had been preserved in the form of originals, copies or transcripts were gathered together into a new Frege archive. After the completion of the editorial work this archive will be handed over to the owner of the original material, the Institute for Mathematical Logic and Foundation Studies of the University of Münster, Westphalia.

On the question whether there were any scientific notes or letters in the possession of Alfred Frege, Frege's adoptive son, which Scholz may not have obtained at the time and which may have survived the war,[7] Christian Frege, an engineer who lives in Bonn, wrote the following in a letter dated 30.1.1972: 'My cousin Alfred Frege – cousin is not quite right: the relationship was much more distant – lived at the time in Berlin W30, Eisenacher St 90–91. This house was completely destroyed by bombing on 22 November 1943; not a thing was saved. Alfred Frege himself was called up to serve at the Torpedo Arsenal West near Paris, where he fell on 15 June 1944.'[8]

On the whereabouts of Gottlob Frege's personal possessions Christian Frege was also able to report (in a letter dated 28.11.1974) that G. Frege, who last resided in Bad Kleinen, Mecklenburg, had – presumably towards the end of his life – bought a house under construction in New Pastow near Rostock. This house bore the designation 'Cottage No. 13'. From the fact that Frege died in Bad Kleinen, Christian Frege concludes that Frege did not live to see the 'completion of the house'. Alfred Frege inherited the house. In the will, Frege's long-time housekeeper, Meta Arndt, was granted the right to live in the house. She died on 8.1.1943. Christian Frege goes on to say about the remaining possessions: 'To my knowledge all of Prof. Frege's remains were transferred from Bad Kleinen to the house in New Pastow.' Before the end of the war Christian Frege paid two brief visits to New Pastow. His description of what he found there will be reproduced verbatim here:

The ground floor of the house, the outbuildings, and the lands were leased to a locksmith, Hünerhoff; my cousin had reserved the two rooms on the upper floor to himself. These were filled with furniture, books, and other household goods, evidently from among Professor Frege's remains. It was *reported* to me [editors' emphasis] that later on, as the Red Army moved in, the house was repeatedly plundered, that it was confiscated for a while to serve as soldiers' quarters, and that it was then filled up to the

---

7 Loc. cit., p. XXXIX.
8 According to information provided by the Chief of Police of Berlin on 23.9.1978, Alfred Frege fell in Montesson near Paris.

roof with refugees – twelve persons in all. I fear that under these circumstances there will hardly be anything left of the remains.

L. Kreiser on the other hand (in a letter dated 23.10.1974) writes the following about the fate of Frege's personal possessions in those days in May 1945:

We now have direct eye-witness reports of what happened in that house in those days, which establish the following fact beyond doubt: the two rooms of the house in which Frege's remains were kept – furniture, books, and sealed boxes – were cleared out a few weeks before the end of the war. As a female eye-witness knows for sure, who was living in Frege's house at the time, the remains were removed under the direction of a relative of G. Frege's housekeeper, Mrs Meta Arndt. That relative was a teacher from Rostock who was living in retirement even then. He has died in the meantime, and so all that remains is the hope that we may learn more about his relations if any, which is what we are now looking for.

The two sources do not contradict each other, if we disregard the fact that Christian Frege did not know that Frege's possessions were removed shortly before the end of the war. Rather, the *report* of which Christian Frege speaks merely seems to refer to a later time.

In a letter dated 8.1.1975 L. Kreiser wrote that he had been once more to Rostock in December 1974 and that he had spoken there with the locksmith Hünerhoff's wife, who was still alive. According to Kreiser, this woman remembers for sure 'that already in the beginning of the 1940's Frege's adoptive son, in uniform and in the company of an elderly man, had cleared a few boxes out of the two rooms, that is, to be more precise, taken them out.' The woman was unable to say anything about the contents of the boxes or their whereabouts. Up to now one can only speculate about the whereabouts of the boxes and the identity of A. Frege's companion.[9]

G. G.   H. H.   Fr. K.   Chr. Th.   A. V.

---

9 Cf. A. Veraart, op. cit.

PREFACE TO THE ENGLISH EDITION

Printed above is the German editors' history of the efforts that have been made to preserve Frege's scientific correspondence. On the whole we are fortunate that so much has survived. It would be unphilosophical, and perhaps improbable, to suppose that what we have lost must have been more interesting than what remains to us. Still, many will mourn the correspondence with Löwenheim, where only the dates of letters, and sometimes not even they, are known to us, and that with Wittgenstein, where we have tantalizingly jejune notes by Scholz on its content. Dates and notes are printed in the German edition for what light they throw on personal or intellectual history. We here give or translate only actual surviving texts, and of those only (but we think all) the ones of scientific interest. Requests for an offprint, refusals to print an article, apologies for not writing – laundry-lists, in a word – we have omitted.

The letters from P. E. B. Jourdain were in English: we give them exactly as they came from his pen, not translating any words or titles of writings in German. The other letters are translated in full, and for titles we follow the general principles of our translation, though in editorial matter we refer to the most accessible English translation of a work by the title it happens to have been given.

The bulk of the letters (including those from B. Russell) were in German. We indicate the few that were in French (or that variety of French once used by Italian scholars) and the one that was in Italian. The formulae of politeness in these languages we replace by what we conceive to have been their contemporary English equivalents.

The letters are arranged according to the alphabetical place of the correspondent other than Frege and then by date. The resulting reference numbers are followed, in brackets and a different style, by the numbers assigned in the German edition. A chronological table of all letters here printed is given at the end. Each correspondent is introduced by a short note on his life and work during Frege's lifetime. These, and the footnotes, are for the most part taken from the German edition, but have been heavily abridged. Exegetical passages, in particular, have been dropped. Translation is already exegesis, and, for the rest, this is an *editio minor*. Any quite new matter is enclosed in square brackets.

As an appendix we print Jourdain's account of Frege's mathematical doctrines with Frege's own comments, for which most of the letters exchanged between the two were a preparation. Here (as with Jourdain's letters) we give exactly his text (and use his abbreviations, not our own), translating only a few sentences that he left in Frege's original German.

Conceived and executed separately from our publisher's edition of Frege's *Posthumous Writings* (1979), this volume may yet be considered a pendant to it. We have not sought exact uniformity, but a general similarity of presentation seemed appropriate. Even the translation often coincides. For example, we too, render *Bedeutung* by 'meaning' (though we prefer 'proposition' to 'sentence' for *Satz*). Both of our renderings are accepted by Frege in his correspondence with Jourdain. Perhaps also we may achieve a similar effect to that of the earlier volume. To see Frege's work in draft or through the eyes of contemporaries may remove the last remnants of that air of a Sinaitic dispensation that has sometimes surrounded it, and may assist the fruitful process of thinking through Frege's problems again which has begun with Professor Dummett's volume[1] and Dr Schirn's collection.[2]

Above I speak in the plural not simply by editorial prerogative but because, while I have attempted to advise Mr Kaal on points of translation, he has given invaluable assistance in editorial matters. All the same, I myself remain responsible for the selection, the abridgement, the general strategy of translation, and indeed for the preface. On specialized points, or ones that seemed such to me, I have been helped by Dr Ian Maclean, Dr Peter Neumann and Professor George Temple, all Fellows of Queen's College, Oxford, whose motto is *Reginae erunt nutrices tuae*.

B. McG.

---

1 Michael Dummett, *Frege, Philosophy of Language* (London 1973).
2 Matthias Schirn, ed., *Studies on Frege*, 3 vols (Stuttgart 1976).

GOTTLOB FREGE

PHILOSOPHICAL AND MATHEMATICAL
CORRESPONDENCE

I FREGE–BALLUE

Editor's Introduction

L. Eugène Ballue (1863–1938) was professor of mathematics at various secondary schools in France. At the time of the correspondence he was teaching at the Lycée de Saint-Quentin. He wrote to Frege in French.

I/1[ii/1] BALLUE to FREGE 20.1.1895

Saint-Quentin
20 January 1895

Dear Colleague,

I have just read your interesting article on whole numbers in the *Revue de métaphysique et de morale*.[1] I must tell you frankly that long before I raised this question in the *Revue* if only in a succinct form (for I had no other aim than to facilitate the understanding of Mr Riquier's article) I never tried to hide from myself the difficulties presented by the nature of whole numbers. The controversy to which my article has given rise and in which Messrs Poincaré, Lechalas, and you yourself, my dear Colleague, have kindly taken part has not diminished my interest in this question; on the contrary, I can only derive satisfaction from seeing competent people discuss my ideas, whether they try to refute them or to render them more precise. Your efforts, my dear Colleague, bear mainly on the latter point, and I can only thank you for it. But probably because of this lack of precision in terms, you make me a supporter of ideas which are not mine. You seem to believe that I attach excessive importance to symbols; yet my article was written mainly as a protest against the introduction into mathematics of a new theory, sponsored by the most brilliant mathematicians, which would have one believe that analysis is nothing but a construction of pure reason, independent of all data of sense, whereas in my estimation it is, on the contrary, impossible to build mathematics on foundations which are not borrowed from experience. In support of my thesis I first adopted the theory currently being advocated and tried to show that in certain proofs it was impossible to do without the consideration of *objects*. You will see that we are close to being in agreement.

I should incidentally be very happy, my dear Colleague, to gain enlightenment about whole numbers by reading your two works. Would you therefore be so kind as to let me have them? I will study them with care and,

---

I/1. 1 Ballue is referring to his article, 'Le nombre entier considéré comme fondement de l'analyse mathématique', *Revue de métaphysique et de morale* 2 (1894), pp. 317–28, and to Frege's reply, 'Le nombre entier' ⟨12⟩.

if you should find this agreeable, I will present and discuss the ideas contained in them for the readers of the *Revue de métaphysique et de morale.*

Thanking you in advance, I am,

Yours sincerely,
E. Ballue
Professor at the Lycée

I/2 [ii/2]   BALLUE to FREGE   9.2.1895

Saint-Quentin
9 February 1895

Dear Colleague,

I have received the *Foundations*,[1] the *Basic Laws*,[2] and the five short works which you were good enough to send me. I thank you for them, and I will set to work at once. It will be a long business, for I neither read nor write German with the facility with which you read and write French, and besides I am very busy at the moment. But with time and effort I shall get there, and I shall be very happy if I can be of service both to you, my dear Colleague, and to science.

Thank you for kindly summing up in a few words the ideas you explained in more detail in your works. This short summary will make my task much easier. It has already given me new ideas, and I believe I understand why the foundations of arithmetic are objective without being real. This thought had already occurred to me, though without its present precision, and I would have expressed it as follows: 'Arithmetical units are not in their nature objects, but nevertheless play the part of objects. They could be called fictitious objects.' I leave you with this thought for what it is worth, that is to say, with a simple sketch, with the embryo of a thought rather than a thought. But I believe it agrees with your conception.

I will only add that since you authorize me to do so I shall not hesitate to ask you for more extensive explanations in the future on any part of your works which may give me trouble.

Yours sincerely,
E. Ballue
Professor at the Lycée

I/2. 1 =GLA.
2 =GGA I.

I/3 [ii/3]   BALLUE to FREGE   21.10.1896

Saint-Quentin
21 October 1896

Dear Colleague,

I have received your little booklet[1] and read it through with much interest, the more so as Mr Peano had published a rather long note concerning his system of logical notation in the July issue of the *Intermédiaire des mathématiciens*. The occasion was this: in January, a correspondent writing under the pseudonym of Lausbrachter had raised the following question:

What is the most general question that has been either raised or settled to date in all of the mathematical sciences?

Mr Peano replied to this question in a note which I am transcribing faithfully in case it has not come to your attention:

Not only did Leibniz state the fundamental rules of the infinitesimal calculus and work on a geometrical analysis resembling vector theory; but throughout his life, from his first work to his last letters, he also entertained the project of a *general algebra*, or *a kind of universal language or script in which all the truths of reason could be reduced to a kind of calculus*. Leibniz speaks of this project in the highest terms, saying that 'a man who is neither a prophet nor a prince can never conceive of doing anything greater for the good of the human species'. Thus according to Leibniz, the most general question that has been raised is the construction of this algebra.

After many preliminary studies by several authors, notably Boole and Mr Schröder, the question has now been settled. All the ideas of logic can be expressed by a small number of signs which can be used in reasoning like the signs $+, -, =, \ldots$ in algebra.

In 1889 I published for the first time a small book written entirely in the symbols of mathematical logic. This first work was followed by several other works by different authors. There now exists a society which publishes the *Formulary of Mathematics*, which is to be a collection of all the known propositions on different subjects in the mathematical sciences, together with proofs and historical indications, all of it written in logical symbols. The first volume of the *Formulary* appeared in 1895; the second volume is now being published. Further information may be obtained from me.

G. Peano (Turin)[2]

I/3. 1 $= \langle 15 \rangle$.
2 The quotation is from *Intermédiaire des mathématiciens* 3 (1896), p. 169.

As you see, Mr Peano makes no allusion to your own works. If you think it right, I could send a rather brief note to the *Intermédiaire* (a collection which accepts only brief notes), worded roughly as follows:

The system of logical notation advocated by Mr Peano has recently been the object of rather forceful criticism on the part of Mr Frege, professor at the University of Jena, in a little work entitled 'On the Conceptual Notation of Mr Peano and My Own'. Mr Frege is also the author of a system of logical notation different from that of Mr Peano, which has been applied to mathematics in his work *The Basic Laws of Arithmetic* (Jena 1893).

Ballue

If you like, I could have this note published either as it is or with whatever modifications you indicate to me. It could not fail to attract attention to your works, which are unfortunately very little known in France.

With my thanks for the parcel, I am,

Yours sincerely,
E. Ballue

I/4 [ii/4]   BALLUE to FREGE   3.1.1897

Saint-Quentin
3 January 1897

Dear Colleague,

I have sent the little note I wrote you about, with the corrections you indicated to me, to the *Intermédiaire des mathématiciens*.

In its November issue the *Intermédiaire* announced that it had received a reply to question 719 from Mr Ballue. This reply will be published one of these days, perhaps even in the December issue; but I have not yet received this issue, which is probably late, as happens rather often with the *Intermédiaire*. As soon as I see that my note has been inserted, I will let you know.

Could you give me some information? Where did Helmholtz publish his paper 'Counting and Measuring', and how could I get hold of it?[1]

With my thanks in advance and my best wishes for the new year, I am,

Yours sincerely,
E. Ballue
Professor at the Lycée

I/4. 1 The reference is to Hermann von Helmholtz, 'Zählen und Messen erkenntnistheoretisch betrachtet', in *Philosophische Aufsätze, Eduard Zeller zu seinem fünfzigjährigen Doctor-Jubiläum gewidmet* (Leipzig 1887), pp. 17–52 [E.T. by M. F. Lowe ('Numbering and Measuring') in H. von Helmholtz, *Epistemological Writings* (Dordrecht and Boston, Mass. 1977).]

I/5 [ii/5]   BALLUE to FREGE   15.4.1897

Saint-Quentin
15 April 1897

Dear Colleague,

The *Intermédiaire des mathématiciens* has finally published in its March issue of 1897, which was only sent to me yesterday, 14 April, the little note of which I sent you the text concerning Mr Peano's system of logical notation.[1] This note was followed by these few lines:

It seems to me that the most general question of mathematics was raised by Mr Peano, the Newton of mathematical logic, and partly settled by the *Formulary*, which is still in progress.

Italo Zignago (Genoa)

I do not know Mr Peano, and I have only a very imperfect idea of his works; but what I know of them seems to me to provide very little justification for the hyperbolical epithet 'the Newton of mathematical logic'. To be sure, this is happening in Italy!

Thinking that this communication will interest you, I am,

Yours sincerely,
E. Ballue
Professor at the Lycée

N.B. Acting on the information you were good enough to communicate to me, I sent for the *Philosophical Essays* dedicated to E. Zeller, and the articles by Helmholtz and Kronecker were of much interest to me.[2]

I/5. 1 The note appeared with only minor changes in *Intermédiaire des mathématiciens* 4 (1897), pp. 66ff.
2 For the reference to Helmholtz, see I/4, note 1. The reference to Kronecker is to Leopold Kronecker, 'Ueber den Zahlbegriff', ibid., pp. 263–74.

II FREGE–COUTURAT

Editor's Introduction

Louis Couturat (1868–1915) taught philosophy at the Universities of Toulouse and Caen as well as at the Collège de France in Paris. His best-known works are *La logique de Leibniz d'après des documents inédits* and *L'Algèbre de la logique*, the latter translated into several languages. He wrote to Frege in French.

II/1 [vii/1]   Couturat to Frege   1.7.1899

Caen
1 July 1899

Sir,

I have the honour to inform you that an International Congress of Philosophy will be held in Paris from 2 to 7 August 1900 and, in the name of the Organizing Committee of which I am a member, to invite you to attend the Congress. The promoters of the Congress, anxious to bring philosophy and the sciences closer together, have given much space on the programme to the Logic of the Sciences and the History of the Sciences. You will find enclosed the list of questions comprised under these two heads. We hope that you will kindly honour the Congress with a report or a communication on those questions that are of most interest to you, as for example on Logic (which is the subject of your *Conceptual Notation*) and on the idea of number, on which you have a theory of your own. We urge you strongly to let the philosophical public profit from your reflections on those subjects on which you have acquired a special competence. We are summoning all those scientists to this gathering who have shown a philosophical and critical spirit in the study of their special science.

In any case, even if you should be unable to attend the Congress, I have been charged officially with asking you to kindly join the Committee of Sponsors of the Congress, which includes already some of the leading figures of French science and philosophy, among whom it will perhaps suffice to mention Mr H. Poincaré, the famous mathematician, who will take an active part in the proceedings of the Congress, and Mr Jules Tannery, whose didactic works impregnated with the rigorous spirit of modern mathematics you will undoubtedly know.

On the other hand we are inviting the principal philosophers and scientists from abroad (notably Germany) and the authors of systems of algorithmic logic, such as Messrs Schröder, Peano, and MacColl; we hope that this subject in particular will give rise to some interesting discussions.*

---

* I recently read your discussion with Mr G. Peano (with whom I correspond quite frequently) in the *Revue de mathématiques*, vol. VI.[1]

II/1. 1 =⟨14⟩.

I shall ask you to kindly communicate the present notice to those of your colleagues to whom it might be of interest and to indicate to me those among them (philosophers or scientists) whom it would be proper to invite to the Congress.

I shall be obliged to you if you let me have your reply as soon as possible, as the Committee of Sponsors and the list of speakers must be settled definitely by the end of this month.

> Yours very truly,
> Louis Couturat
> Professor of Philosophy at the University of Caen

P.S. The principal organizers of the Congress (under the auspices of the Minister of Public Instruction) are my friends from the *Revue de métaphysique*, which you honoured with a communication in 1895.[2] I must excuse myself for not having mentioned your works and your theory in my thesis *On Mathematical Infinity* (1896, but finished in 1894) because I knew them only through Mr Husserl's *Philosophy of Arithmetic*, which I read while I was completing my work.

II/2 [vii/2]   COUTURAT to FREGE   8.7.1899

> Caen
> 8 July 1899

Sir,

I thank you warmly, in the name of the Organizing Committee, for the honour you have done us by agreeing to join the Committee of Sponsors of the Congress of Philosophy. We regret that you think you may not be able to attend, all the more because, as you are kind enough to say, the Congress will deal with certain questions that are of interest to you and on which you would undoubtedly have original and instructive ideas to express. Let us hope that you *will* be able to honour us with your presence and take part in our discussions.

You will receive shortly a complete (printed) programme of the Congress, and we should be obliged to you if you communicated it to anyone who might be interested in it whenever you have the opportunity. Since the list of speakers must be closed in November, announcements of communications will be received up to the end of October.

With our sincere thanks, I am,

> Yours very truly,
> Louis Couturat

---

2 = ⟨12⟩.

P.S. Here is the list of philosophers and scientists from abroad who are already members of the Committee of Sponsors:

Germany: Messrs Zeller   Eucken
                Paulsen   Frege
                Riehl   Haeckel
                Vaihinger   Deussen
                Kuno Fischer   Ostwald
                Moritz Cantor   Natorp

Austria: Mach
Switzerland: Stein
Denmark: Høffding
Holland: Korteweg
Italy: Labriola, Cantoni, Peano
England: Leth, Ritchie
United States: Baldwin
(Some replies have not yet been received.)

II/3 [vii/3]   COUTURAT to FREGE   6.1.1901

Paris
6 January 1901

Sir,

I am very pleased that my article on the Congress of Philosophy should have been of interest to you;[1] it is hardly more than a partial summary of the proceedings published in the *Revue de métaphysique*, which you must have received and whose Section III was written by me.[2] In any case, your questions evidently refer to the latter. I hasten to give you the answers I can at the moment; if you find them insufficient, you can ask me new questions, and I will pass them on, if necessary, to the authors concerned.

The first question can be answered as follows: *A always* designates the affirmation of proposition *A*, or more exactly, its domain of validity. Likewise, *A + B* designates not only a thought but an *affirmation*, the alternation of *A* and *B*, whose domain of validity is the sum of the domains of *A* and *B*. There is therefore no inconsistency here in the notation.

Concerning the second question: Mr MacColl's double division (τ true, ι false) and (ε certain, θ variable, η impossible) applies to the same set of propositions, i.e., to all *general* propositions which bear on a multitude of particular cases and which can be true in some cases and false in others.

II/3. 1 Couturat is referring to his note, 'Les mathématiques au Congrès de Philosophie' in *L'Enseignement mathématique* 2 (1900), pp. 397–410.
2 The reference is to 'Congrès International de Philosophie', *Revue de métaphysique et de morale* 8 (1900), pp. 503–698. The questions discussed below refer to pp. 563ff.

Certain means always true; impossible = always false. True propositions include certain propositions, false propositions include impossible propositions; and variable propositions can be divided, in each given case, into true and false ones. Here is the schema by which Mr MacColl represents these relations:

$\tau\varepsilon$	$\tau\theta$
$\iota\eta$	$\iota\theta$

It goes without saying that for singular propositions there are only two possible cases, τ and ι, which then coincide with ε and η. Thus the proposition $3 + 4 = 7$ is τ and ε, the proposition $3 + 4 = 25$ is ι and η.[3]

Third question: Mr MacColl's implication $A : B$ (like Mr Schröder's $A \in B$, Peirce's $A \prec B$, etc.) in no way presupposes the truth of $A : A$ can be false and B true. In general, the false (symbolized by zero) can be the antecedent of any implication whatever, i.e., we always have $0 \prec x$, whatever x may be.

As for complete induction, which Mr Padoa turns into an axiom and then uses to *define* a whole number, this theory is too long to be explained here. It is explained in Mr Peano's *Formulaire de Mathématiques* and in his previous works.[4] There you will also find a proof of the fact that this axion cannot be logically deduced from the other axioms of arithmetic.

So that you can criticize these theories and discuss them at greater leisure, I advise you to await the publication of the full text of the papers read at the Congress of Philosophy: the third volume will appear in about three months (under my care). The papers in question are at the printer's.

Mr MacColl should not be taken to task for ignoring your *Conceptual Notation*, for he does not read German. I do not have the same excuse: I own your book, and I will read it when I study algorithmic logic, on which I expect to publish a work. But at the moment I am absorbed in other work which I am hard pressed to finish. This is why I shall ask you to excuse the brevity of my reply. I hope this will not prevent you from communicating to me the reflections which the papers read at the Congress of Philosophy may suggest to you. I should also be very pleased to have your opinion on the question of an *international language,* a project delegated to me by the Congress and about which I have said a few words in the postscript to my article. I am preparing a circular letter on this subject which I will send you.

---

3 Cf. MacColl, 'La logique symbolique et ses applications', *Bibliothèque du Congrès International de Philosophie* 3 (1901), pp. 135–83.
4 Cf. *Formulaire de Mathématiques* III (Paris 1901), pp. 41ff.

In the meantime I thank you for your kind wishes and offer my sincerest wishes for you personally and for your work.

Yours very truly,
Louis Couturat

II/4 [vii/4]   COUTURAT to FREGE   18.6.1901

Paris
18 June 1901

Sir,

Please excuse me if my ever more absorbing tasks have not left me the leisure to reply *with due reflection* to your last letter. Only today do I find the time to reflect on the questions you raise and the solution to the difficulties you mention.

On the first point, it is clear that if 'A' in isolation signifies the affirmation of proposition A, it no longer has this sense of absolute affirmation in the alternation $A + B$ and even less so in $(A + B) + C$ or in $(A + B)C$. But I believe I can escape this difficulty by regarding A, B, $A + B$, etc. as representing, properly speaking, the domains of validity of the propositions mentioned. Likewise, a, b, $a + b$, etc. in the calculus of concepts represent not the concepts *themselves* (i.e. their content) but the corresponding *classes* (i.e. their extension). This is why the two calculi (or the two classes of propositions) coincide formally: they are one and the same calculus, a calculus of *sets* in the mathematical sense of the word.

On the second point (Mr MacColl's variable judgements) my explanations were poor, i.e. too brief. I said that they are *general* judgements: this must not be taken to mean *universal* judgements (All A are B) which are absolutely determined and which are consequently (always) true or (always) false, but rather indeterminate or, as you say, incomplete judgements, as illustrated by the example 'It is raining'. These are judgements that contain a variable or indeterminate term x, where this x can take all the values of a certain set, i.e., represent each individual or *point* of a given class. Now all the judgements considered in the theory of probability are of this kind. Their *probability* is the relation of the number of cases in which they are *true* to the number of cases for which they have a sense (or to the number of all the given x's). Hence, one can call them *variable* or better *variable in sense*, and represent their value by a fraction intermediate between 0 and 1. In any case, to acquaint you better with Mr MacColl's system, I am going to ask him to send you an offprint of the paper he read at the Congress of Philosophy, which sums up his entire work.[1] I hope he will

---

II/4. 1 The reference is to MacColl, 'La logique symbolique et ses applications', *Bibliothèque du Congrès International de Philosophie* 3 (1901), pp. 135–83.

be able to meet my request and that you in turn will be able to write a sympathetic critique of this paper, for which both the author and I myself (as editor) would be grateful.

Your last letter seems to indicate that you had yet further things to tell me; please communicate them to me now that I have a little more time.

There is a subject on which I should be happy to have your opinion: the one defined by the letterhead of this letter[2] and discussed in a little booklet I wrote and sent to you recently.[3] Our enterprise appears to meet a practical need which is becoming more and more urgent, for it has promptly received the support of very important societies like the Touring-Club. It has had at least as good a reception in the scientific world, not only in France but also in Germany, Italy, Austria and Belgium, and our publicity is only beginning. This is a question on which opinions are very much divided; but we are satisfied if a person subscribes to the two fundamental points of our Declaration: (1) the utility and (2) the possibility of an international language.

I look forward to hearing from you.

> Yours very truly,
> Louis Couturat

II/5 [vii/5]   COUTURAT to FREGE   13.10.1901

> Paris
> 13 October 1901

Sir,

I thank you warmly for your last letter, which deals with the two topics to my complete satisfaction. On the international language I have nothing to add to your judicious remarks, and I confine myself to stating that I am altogether in agreement with you: it is for scientific, commercial and utilitarian purposes that we require an international language, in a domain where the expression of *facts* and *ideas* must *not* be invested with a national character, and where it is already *in fact* largely international. As for the only difficulty you foresee, that of indispensable *neologisms*, the same difficulty exists in all the national languages and will be resolved in the same manner. There is nothing to prevent the institution of an *academy* of the international language to sanction new words and to maintain the unity and purity of the language. In the absence of linguistic *intuitions*, we will have absolutely fixed rules of logic and analogy to guide us in the formation of

---

2 The letterhead reads: 'Delegation for the Adoption of an Auxiliary International Language'.
3 Cf. L. Couturat and L. Leau, *Histoire de la langue universelle* (Paris 1903).

words; and it is only in exceptional cases where these rules leave us with some hesitation that we shall have recourse to the Academy. The latter (like the French Academy) will only have to sanction and regularize *usage*. As far as translations are concerned, the difficulties will be the same as from one national language to another, and *even less*. As always, literary and poetical works will lose the most; to savour them fully one will always have to learn the original language. All the international language can do is to place ancient and foreign works within reach of the general public *as best it can*, just as translations into French for example acquaint us with Ibsen and Tolstoy, or even with Shakespeare and Dante, Homer and Virgil, for those who cannot read the original (and they constitute the *vast majority*).

On the question of logic I accept your explanations: one is of course free not to admit the concept, which is at least paradoxical, of an 'incomplete' or 'indeterminate thought' or 'judgement'. But as you yourself indicate, one must then also abstain from using the notion and the expression 'indeterminate number' and all others of the same kind. I acknowledge that this makes for greater rigour and clarity; but I wonder whether this would be very convenient in practice: one would have to reform the whole language of the mathematicians. Perhaps one would then discover (as one almost always does) that the expressions one is trying to proscribe have their *utility* in that they correspond if not to a *real* then to some *ideal* object. In short, you transfer the indeterminacy from the judgement to the concept: you say that 'the number of cases where a judgement is true' is in reality 'the number of objects subsumed under a concept'. By pushing the consequences of your theory to their limits, one might perhaps come to banish the concept itself, as a *general* representation of an *indeterminate* object, i.e., as an *incomplete* (abstract) representation. This would be pure nominalism. I will not undertake to discuss this here. I only want to state that we can *in fact* subsume several different objects under one and the same concept or, so as not to prejudge anything, under one and the same *rubric, x, y, ...* Hence, if one admits such rubrics (symbols for sets of well-defined objects), I see no sufficient reason for rejecting judgements of the form '*x* is *a*', whether valid for the totality of *x*'s (i.e., certain judgements) or only for a certain portion of *x*'s (i.e., probable judgements). To tell the truth, I believe that this is no more than a question of words or terminology. As for the rest, I acknowledge freely that the domain of validity of a judgement is a set like any other, and it is for this very reason that judgements are subject to the same formal calculus as concepts (considered as classes of objects).

This to me is the fundamental idea of algorithmic logic, an idea Mr Schröder did not sufficiently bring to light.

Since you are favourably disposed to our enterprise, I conclude by asking you to make it known to your circle and to commend it to the attention of your colleagues and friends. I shall send you (*free of charge*) all the documents you may want, and in any number you like. I am sending you provisionally our circular in German and English as well as the last Report of the Delegation. May I remind you that our Delegation solicits the support of all learned societies as well as the nomination of delegates?

> Yours very truly,
> Louis Couturat

II/6 [vii/6]   COUTURAT to FREGE   11.2.1904

> Paris
> 11 February 1904

Sir,

I have the honour of sending you an offprint of the first of the articles I am going to publish on the principles of mathematics,[1] a series inspired by Mr Russell's book.[2] I should be happy to receive your observations and criticisms on the subject. I must make a confession to you and even offer you my excuses. When I wrote the first article (and even the second, which is about to be printed) I did not yet know your works. I had already obtained your *Conceptual Notation* several years ago: I wanted to study it to give an account of it in my future *historical* work on algorithmic logic; but I had not yet found the time to study it, since the preparation of that work had been interrupted by my work on Leibniz. It was at the instigation of Mr Russell, who praised your works in the highest terms in his letters, that I began studying your works, and I am happy to have done so, for I believe I shall derive much profit from them. To tell the truth, on first approach they are not very inviting, and the symbolism they employ makes them difficult to read. I still cannot boast that I can read them fluently or that I understand them. But what I can nevertheless say even now is that (like Mr Russell) I share entirely your general philosophical opinions, concerning for example the empiricist logic of the psychologists and the so-called formalist logic of the mathematicians, and that I profoundly admire the exactness and rigour of your logical principles. I especially appreciated your *Foundations of Arithmetic* with its considerable philosophical range; and I am counting on it as a source of inspiration for the article I am going to write on Kant and modern mathematics for the May issue of the *Revue de métaphysique*,[3] which is devoted to Kant on the occasion of the centenary of his death. I

II/6. 1 L. Couturat, 'Les principes des mathématiques', *Revue de métaphysique et de morale* 12 (1904), pp. 19–50.
2 B. Russell, *The Principles of Mathematics* I (Cambridge 1903).
3 An article with this title ('Kant et la mathématique moderne' appeared in *Bulletin de la Société française de philosophie* 4 (1904), pp. 125–34. Another article ('La philosophie des mathématiques de Kant') appeared in *Revue de métaphysique et de morale* 12 (1904), pp. 321–83.

drafted this article before I had read your book, and arrived at the same conclusion as you: that arithmetical judgements are analytic while geometrical judgements are synthetic. I shall, of course, use every opportunity to cite your works, both in this article on Kant and in the others.

I am even planning to do more, and since I neither could nor can incorporate your theories in the series of articles I have started to write, I shall devote a separate article to your logico-mathematical theories, stripping them as much as possible of the symbolism which must have discouraged many of your readers, which in any case will certainly discourage mine, and which would even cause difficulties to the printer.

In this way I hope to do my share in vulgarizing (popularizing) your theories among the French philosophical public; I hope I will not seem to you to be vulgarizing them in the bad sense (of rendering them *vulgar*).

I remember having received (either from you or from the *Revue de métaphysique*) your article 'On the Numbers of Mr Schubert',[4] and that I wrote an account of it at the time in the *Revue de métaphysique*.[5] In spite of the great care with which I preserve all the booklets I receive, I have not been able to find it. On the other hand, I have not been able to find the two papers you published in 1885 and 1891 in the *Sitzungsberichte der Jenaischen Gesellschaft* in any library, not even at the Institute.[6] This is all I lack for a knowledge of *all* your works. If you could let me have *or just lend me* these two papers, I should be very grateful to you.

Yours very truly,
Louis Couturat

II/7 [vii/7]   COUTURAT to FREGE   21.10.1906

21 October 1906

Dear Colleague,

Thank you for sending me your articles from the *Jahresbericht*; I have read them with interest, and I have just written a short account of them for the *Revue de métaphysique*.[1] You have no doubt seen Mr Poincaré's articles about (and against) logistic, in which he confuses Mr Hilbert with the logisticians! It would be desirable, in the interest of the diffusion of your ideas, if you were to write a *general* and *very brief* exposition of your logical

---

4 =⟨16⟩.
5 *Revue de métaphysique et de morale* 8 (1900), March Supplement, pp. 4f.
6 ⟨6⟩ and ⟨7⟩ are meant.
II/7. 1 *Revue de métaphysique et de morale* 15 (1907), January Supplement, pp. 11f.

theory. But perhaps you will find this difficult. This is what I will try to do in any case in my *History of Logistic*, on which I am working at present and in which one chapter will be devoted to you.[2]

Yours sincerely,
Louis Couturat

---

2 The reference is presumably to a version of a course of lectures on the History of Modern Formal Logic which Couturat gave at the Collège de France in 1905–6. None of this was published, except for the introductory lecture which appeared under the title 'La Logique et la philosophie contemporaine' in *Revue de métaphysique et de morale* 14 (1906), pp. 318–41.

Editor's Introduction

Hugo Dingler (1881–1954) graduated from Munich in 1906 in mathematics, physics, and astronomy. After qualifying as a university lecturer in 1912 and taking part in World War One, he was named reader at Munich University in 1920. In 1932 he was appointed professor of philosophy at the College of Science and Technology in Darmstadt, but had to leave this chair in 1934 for political reasons.

III/1 [ix/2]   FREGE to DINGLER   31.1.1917

Jena

31 January 1917

My dear Doctor,

Thank you very much for sending me the two publications.[1] I myself have also given some thought to sets. You know perhaps that I diverge in some respects from Hilbert's opinions, and you will not therefore find it surprising if now and then I also take exception to your position. This should not deter us from trying to reach an understanding. For to a scientific author a conflicting opinion, if carefully stated and well-founded, must always be preferable to an attitude of indifference. For there is something he can hope to learn from such a conflict, if only how he can better guard his own opinion against misunderstandings by expressing himself more precisely. Therefore you will not, I think, be offended if I indicate to you the places to which I take exception. I will first look at your proposition II/4 ('If we succeed ...').[2] Is this case at all possible? If we derive a proposition from true propositions according to an unexceptionable inference procedure, then the proposition is true. Now since at most one of two mutually contradictory propositions can be true, it is impossible to infer mutually contradictory propositions from a group of true propositions in a logically unexceptionable way. On the other hand, we can only infer something from true propositions. Thus if a group of propositions contains a proposition whose truth is not yet known, or which is certainly false, then this proposition cannot be used for making inferences. If we want to draw conclusions from the propositions of a group, we must first exclude all

---

III/1. 1 As shown by the quotation below (cf. note 2), one of the two is an offprint of Dingler's essay, 'Über die logischen Paradoxien der Mengenlehre und eine paradoxienfreie Mengendefinition', *Jahresberichte der deutschen Mathematiker-Vereinigung* 22 (1913), pp. 307–15.

2 Dingler's 'proposition II/4' reads: 'If we succeed in inferring logically from a group of premises that a certain statement both holds and does not hold for one of the concepts contained in the premises, then I say: *This group of premises is contradictory, or contains a contradiction*' (ibid., p. 308).

propositions whose truth is doubtful. The schema of an inference from one premise is this:

'*A* is true, consequently *B* is true.'

We have this case, e.g., where *A* is a general proposition and *B* a special case of it.

The schema of an inference from two premises is this:

'*A* is true, *B* is true, consequently Γ is true'.

It is necessary to recognize the truth of the premises. When we infer, we recognize a truth on the basis of other previously recognized truths according to a logical law. Suppose we have arbitrarily formed the propositions

'2 < 1'

and

'If something is smaller than 1, then it is greater than 2'

without knowing whether these propositions are true. We could derive

'2 > 2'

from them in a purely formal way; but this would not be an inference because the truth of the premises is lacking. And the truth of the conclusion is no better grounded by means of this pseudo-inference than without it. And this procedure would be useless for the recognition of any truths. So I do not believe that your case II/4 could occur at all.

Yours sincerely,
G. Frege

III/2 [ix/3]   DINGLER to FREGE   2.2.1917

Augsburg
2 February 1917

My dear Professor,

Let me first thank you for your lines of 31.1.17. They gave me much pleasure, especially since they came from a man whose activities seem to have done much to bring the problem of the 'foundations' of mathematics home to the mathematicians, even though I am also conscious of differing, as far as I can tell without further discussion, on some fundamental points. I completely agree with you that a conflicting opinion, if serious and well-founded, can only be welcome, for in many cases it furthers what is the common goal of us all: truth and knowledge. I would now like to go into the scientific substance of your valuable letter, though I should mention beforehand that I would certainly have produced more and done more to analyse and elucidate my position if this war had not intervened and kept me busy for more than two years already with the heavy demands of military service and without even a leave to speak of.

The principal difference between us goes very deep: I believe that it is unnecessary and even inadmissible to bring the concept of truth into mathematics as you propose to do in your letter. You write: 'On the other hand, we can only infer something from true propositions.' It seems to me that the 'truth' of the premises is completely irrelevant to the validity of the inference. We seem compelled to talk about the truth of the premises when we make use of the logical calculus of truths. However, in mathematics the pure calculus of concepts seems to me to be perfectly sufficient in the majority of cases, so that the application of the calculus of truths represents at best a needless complication. But it would take me too far to pursue this point.

The natural course of things would seem to be this: I create a new concept by synthesis, say the concept of a plane finite in both dimensions and with a constant negative curvature. Hilbert has shown that such a plane is impossible.[1] But if one does not know this, even the most expert mathematician cannot tell by simply looking at this concept that it is contradictory. If I now apply to this concept the propositions of plane geometry that have been proved previously and inquire into the concept's properties, then I can, if I follow the proper procedures, obtain two propositions about the concept which are opposed to each other as contradictories. But in all this nothing has been said, nor could anything be said, about the 'truth' of the statements about the concept I have created, which are contained in its definition. I here mean truth in the metaphysical sense. If I want to draw further conclusions from a group of statements, it seems to me natural to presuppose (ad hoc) that they are 'true', but there would be some equivocation here, for 'true' in this case means merely 'chosen as a premise for my reasoning of the moment'.

But this can also be proved somewhat more formally. If I create a concept like that of a plane finite in both dimensions and with a constant negative curvature, then it becomes the subject of my further reasonings and must therefore be given a name if I do not want to repeat its entire definition each time. If I call it X, the premises of my reasoning are:

(1) X is an area with a constant negative curvature.
(2) X is finite in both dimensions.

Now it can be seen immediately by looking at these propositions that they can be neither true nor false (in the metaphysical sense), but that they are merely hollow forms, as it were, to be filled later by real propositions. For

---

III/2. 1 Cf. D. Hilbert, 'Über Flächen von konstanter Gaußscher Krümmung', *Transactions of the American Mathematical Society* 2 (1901), pp. 87–99, which is reprinted in D. Hilbert, *Gesammelte Abhandlungen* II (1933, reprinted 1965), pp. 437–48.

they contain a concept, X, which is not yet present and which is still hardly known. And what is at issue here is not whether the two propositions are true in themselves but whether they are 'true' *together*, i.e., whether or not they are contradictory.

So this seems to me to be the crux of our difference, judging by your valuable letter. As far as my philosophical concept of truth is concerned, it can probably best be gathered from my *Foundations of the Philosophy of Nature* (Leipzig 1913),[2] and I have dealt with my conception of the nature of mathematical reasoning, though unfortunately only in very little detail, in a recent work on *The Principle of Logical Independence in Mathematics: an Introduction to the Axiomatic Method* (Munich 1915).[3]

This is roughly the position which I believe I have achieved after some struggle and much vacillation. Your position is nonetheless of great interest to me, and I find your stimulating exposition of it very valuable. I understand that few have as much right as you to scrutinize a solution to Russell's paradox. My work represents, as it were, only the skeleton of a line of thought which is still in need of more penetrating and more detailed axiomatic analysis.[4] But one cannot do everything at once, and this enterprise is very long and complicated. But I do not believe that my solution can in the long run be shown to be fundamentally mistaken; I believe that as time passes it will more and more turn out to be correct in its basic features.

Once more, many thanks for the pleasure you have given me with your lines. Perhaps we can still come a little closer together in our views.

Yours sincerely,
H. Dingler

III/3 [ix/4] FREGE to DINGLER 6.2.1917

Jena
6 February 1917

My dear Doctor,

Many thanks for your detailed letter! I wish I had succeeded in bringing the problems of the foundations home to the mathematicians. Up to now, however, I see little evidence of success. With regard to the substance of your letter, I believe that we must first come to an understanding about the

---

2 *Grundlagen der Naturphilosophie* (reprinted Darmstadt 1967).
3 *Das Prinzip der logischen Unabhängigkeit in der Mathematik, zugleich als Einführung in die Axiomatik.*
4 The reference is probably to the essay, 'Über die logischen Paradoxien der Mengenlehre und eine paradoxienfreie Mengendefinition', *Jahresberichte der deutschen Mathematiker-Vereinigung* 22 (1913), pp. 307–15.

expressions 'proposition' and 'premise'. If I am not mistaken, logicians distinguish a proposition as something external, audible, and visible from a judgement, the content expressed by the proposition. So logicians regard a proposition as the less essential of the two and take it into account only because of the content it expresses. On the whole, mathematicians draw a less sharp distinction. The proposition and its contents may sometimes get blurred in their minds, and to be concerned with its sense may appear to many as unworthy of a mathematician, even though they too must be aware, if only very obscurely, that on the deepest level the sense or content is the important thing. Terminologically I side more with the logicians when I take a 'proposition' to mean . . . and distinguish it from a thought, as its sense or content.[1] According to my way of speaking, we think by grasping a thought, we judge by recognizing a thought as true, and we assert by making a judgement known. It is one thing merely to express a thought and another simultaneously to assert it. We can often tell only from the external circumstances which of the two things is being done. What an actor says on the stage has usually the form of an assertoric proposition and would also be understood as an assertion if it was said off-stage; but we know that on stage it is not said in earnest, but only playfully. The actor only acts as if he were asserting something, just as he only acts as if he wanted to stab someone, and he cannot be charged with lying any more than with attempted murder. What is spoken on stage is said without assertoric force. But in the language of science, too, a thought is sometimes merely expressed without being put forward as true, e.g., in interrogative sentences or conditional clauses. This is why I distinguish between thoughts and judgements, expressions of thought and assertions. Let me go on to complexes of propositions. The most important case in mathematics is probably the case of a propositional complex consisting of an antecedent and a consequent. It might at first be thought that the antecedent expresses a thought, that the consequent expresses a thought, and that the whole propositional complex expresses a thought. But the situation is such that we can recognize the truth of the last thought without thereby recognizing the truth of the first two propositions. Thus we can recognize the proposition 'If $13^{13} > 23^{11}$, then $13^{13} + 1 > 23^{11} + 1$' without knowing whether $13^{13} > 23^{11}$ and whether $13^{13} + 1 > 23^{11} + 1$. From the proposition '$13^{13} > 23^{11}$' or better from its thought content we cannot infer anything as long as we do not know that it is true. I cannot recognize different meanings of the word 'true'. We can indeed prove the proposition 'If a first number is greater than a second, then the first number augmented by 1 is greater than the second number augmented by 1' or 'If $a > b$, then $a + 1 > b + 1$', or in the symbols of my conceptual notation:

$$\vdash \begin{array}{l} a+1 > b+1 \\ a > b \end{array}$$

III/3. 1 This sentence is incomplete in the original text.

This proposition seems to be of exactly the same kind as the proposition expressed above or as the simpler proposition 'If $3 > 2$, then $3 + 1 > 2 + 1$', or in my symbols:

$$\vdash \begin{array}{l} 3+1 > 2+1 \\ 3 > 2 \end{array}$$

and yet the case is entirely different. After we have recognized the proposition '$3 > 2$' as true, we can use it to prove the proposition '$3 + 1 > 2 + 1$'; but we cannot use the proposition '$a > b$' to prove the porposition '$a + 1 > b + 1$'; for '$a > b$' is not a proper proposition because it does not express a thought, nor consequently can it be recognized as true. The proposition

$$\vdash \begin{array}{l} a+1 > b+1 \\ a > b \end{array}$$

does not express three thoughts but only a single one. The letters 'a' and 'b' do not designate anything but only serve to express a general thought. I call '$a > b$' an improper proposition because it has the form of a proposition but does not express a thought. After we have recognized the propositions

$$\vdash \begin{array}{l} a+1 > b+1 \\ a > b \end{array}$$

and '$3 > 2$' as true, we can use them as premises in an inference and thus prove the proposition '$3 + 1 > 2 + 1$'. The proposition

$$\vdash \begin{array}{l} a+1 > b+1 \\ a > b \end{array}$$

has a sense only because the 'a' above points to the 'a' below, and conversely. We obtain a proper proposition from an improper one by replacing the letters which merely indicate – or in language, such words as 'something' and 'it' – by signs which designate. It seems to me probable that if we think we can infer something from a proposition which is not true because it does not express a thought, then this apparent premise is an improper proposition which appears linguistically as an antecedent within a propositional complex consisting in addition of a consequent which is likewise an improper proposition and perhaps in addition of other improper antecedents. The whole propositional complex may be a proper proposition expressing a true thought. But the individual improper propositions which make up the propositional complex, when torn out of the context of the whole, are useless and have no sense, even though they may have parts that have sense. We can go from a true general thought expressed in a propositional complex to a special case contained in it by replacing a letter which occurs in it, e.g., 'a', by the same designating sign wherever it occurs in the propositional complex. For the sake of brevity, I speak here of 'the same sign'. In the different places there are of course different things. If these things have the same shape because they are intended to designate the same object, and if this intention is recognizable, then I call these things the same sign, for the sake of brevity and convenience of expression. Thus the propositional complex forms a self-contained whole such that the same designating sign must be put wherever 'a' occurs within the propositional complex, whereas it makes no difference whether this is done outside the propositional complex. Once we have recognized the content of the original propositional complex as true in a judgement, we can also recognize as true the content of a propositional complex arising from it by substitution. We have drawn a conclusion by going from the general to the particular contained in it. For the generality effected by the letter 'a' consists precisely in this: that such a transition is permitted. Now it can happen that, in making that inference, we derive a proper proposition from one of the original improper antecedents and that this proper proposition expresses a true thought. We can then assume the truth gained by that inference as our first premise and the truth of the thought expressed in the proper proposition, which was obtained by substitution and recognized in a judgement, as the second premise of an inference. The conclusion then arises from the first premise through suppression of that part which constitutes the content of the second premise. Example:

$$\vdash \begin{array}{l} a+1 > 2 \\ a > 1 \end{array}$$

This is the original propositional complex. The vertical stroke on the left side of the upper horizontal is the judgement stroke by which the truth of the content is recognized.

$$\vdash \begin{array}{l} 2+1 > 2 \\ 2 > 1 \end{array}$$

This proposition is gained from the original propositional complex by replacing the letter 'a' with the designating sign '2'. This transition corresponds to an inference. It is a transition from the general to a particular contained in it. This proposition constitutes the first premise. From the improper proposition '$a > 1$' we now get the proper proposition '$2 > 1$'. The latter, recognized as true, constitutes the second premise:

$$\vdash 2 > 1$$

The conclusion is

$$\vdash 2+1 > 2$$

and this is the first premise after suppression of that part which contains the thought expressed by the second premise. In the conclusion the second premise has disappeared entirely and the first at least partly. And this entire or partial disappearance of the premises is what is characteristic of a proper inference. The connection is often hidden by the way we proceed. It is said, e.g., 'Let $a > 1$', and then after some reasoning, 'So $a + 1 > 2$'. In this case the path is of course only a short one, but in other cases it can be quite long. And now it is thought that we have proved the proposition '$a + 1 > 2$' from the axiom '$a > 1$' – if we can call it that. This is completely wrong! What has been proved is the content of the proposition 'If something is greater than 1, then what we get by adding 1 to it is greater than 2', or in symbols and more precisely:

$$\vdash \begin{matrix} a+1 > 2 \\ a > 1 \end{matrix}$$

By saying 'Let $a > 1$' one creates the impression of an independent proposition, whereas properly speaking it is only an improper antecedent which has no sense at all by itself but only yields a whole with a sense together with the improper consequent 'So $a + 1 > 2$'. If we want to write down the result of the proof, we must not write '$a + 1 > 2$', for this does not say anything, but we must write:

$$\vdash \begin{matrix} a+1 > 2 \\ a > 1 \end{matrix}$$

and here the apparent premise or, if you like, the apparent axiom '$a > 1$' has not disappeared but is still present, as large as life, which shows that it was only an apparent premise.

Yours sincerely,
G. Frege

III/4 [ix/5]   DINGLER to FREGE   26.2.1917

Augsburg
26 February 1917

My dear Councillor,

Under the pressure of military service I did not get around till today to thanking you for your detailed and substantial letter and to replying to it as far as this is possible within the framework of a letter. You begin very appropriately with a discussion of some basic concepts: proposition and premise. Please let me start at once with the point that seems to be at the centre of the difference between us, as far as the content of your second letter is concerned, as I believe I can formulate my objections most clearly in this way.

If I understand you correctly, there is for you a fundamental difference between a 'proper' and an 'improper' proposition; as your examples you choose the expressions '3 > 2' and '*a* > *b*'. Let me start at once with these examples. I have given a detailed sketch of my view of the general concept of number in the first part of my work, *The Principle of Logical Independence in Mathematics* (1915), which I will call LI for short.[1] For me, '2' and '3' are also mere signs which receive a 'sense' only through the premises or axioms that are assumed to hold for them. The proposition '3 > 2' only seems to have an independent sense, and it seems so when I work directly with the 'imagination', as is the case in the pre-logical or pre-axiomatic stage of science. '3', '>' and '2' are then 'popular concepts', as I called them in my *Foundations of the Philosophy of Nature*,[2] i.e., they are ideas abstracted from practical life with which the understanding operates in an intuitive manner, which is in fact how everyone operates at all times in ordinary life. But this state cannot be the ideal of science, or the form science ought to assume. It seems to me that the 'ideas' someone has when he sees the signs '2' and '3' must not as such be taken as the foundations of science but must be regarded in their logically analysed form. But then the proposition '3 > 2' is not an independent proposition; it is only part – as you so well put it – of a *propositional complex* which rests on a series of irreducible premises or axioms. But then the difference between '*a* > *b*' and '3 > 2' is only one of *degree* and not of kind: in both cases two signs are joined by '>'; only in the one case there are more premises concerning these signs than in the other case. Thus '*a* > *b*' represents only *one* premise, while '3 > 2' carries with it a whole group of premises, its 'logical foundation', as I called it in LI.[3] Without these premises '3 > 2' does not differ in any way from '*a* > *b*', for without them '3' and '2' are completely arbitrary signs, just like any others.

From what I have said you should already be able to form a rough judgement about my position. I am not saying by any means that the logic of the logicians is somehow false; all I mean is that it is mainly adapted to the matter out of which it was created: the reasonings of ancient philosophy as well as of certain branches of modern philosophy and the kind of explanatory description which is now customary in this science as it had to be in the nature of the case. However, as I have often explained in my writings,[4] I regard mathematics as a kind of higher stage in the development of science, though of course without casting aspersions on other sciences. I believe we need an entirely new kind of logic for this kind of synthetic science, one which exists at present only in a rudimentary form, a logic which provides explanations and justifications for the many new phenomena provided by this kind of science. A characteristic and very revealing element

---

III/4. 1 *Das Prinzip der logischen Unabhängigkeit in der Mathematik* (Munich 1915).
2 *Grundlagen der Naturphilosophie* (Leipzig 1913; reprinted Darmstadt 1967), p. 22.
3 op. cit., p. 60.
4 Cf. ibid., pp. 20 ff.

of this new logic seems to me to be the principle of the logical mapping of different sciences onto one another, which I have discussed in many of my writings and which shows, it seems to me, that in *these* parts of science there need be no immediate connection between the propositional form and the 'meaning' except in the irreducible axioms.

Within *this* kind of logic I can call a proposition true only if it belongs to a logical structure that is free from contradiction and has been correctly derived within it. If this logical structure has an interpretation in reality, then this proposition is a 'true proposition' of the science arising from this interpretation. In the non-synthetic sciences there are also certain concepts of truth, but these must be characterized in a more complicated way.

Once more, many thanks for your valuable and stimulating lines.

Yours sincerely,

H. Dingler

[*Appendix*][5]

After this you will understand it if I take the view that much of what is said by logicians coming from the philosophical side is inapplicable to mathematics, since the logic pursued there represents only a kind of preliminary stage to the logic required for the exact sciences. Not that this logic is somehow false; I only believe that it is not the last word. But it is understandable if one tries as far as possible to establish what is theoretically more general and more ideal. I believe that in an exact science, as for example number theory, all axioms whatsoever must ideally be reducible to pure stipulations, and that *everything* else will then be logically derived from them, so that propositions with a 'proper' content, which could not be obtained in this way, will no longer occur within the science. I am only saying these things to give you a brief hint as to how I see these things within a wider context. My position on the concept of probability also derives from this. If a logical structure is free from contradiction, then every proposition correctly derived within it is for me '*logically true*'. If a logical structure has an interpretation in reality, then a proposition so derived is in this science (i.e. in this interpretation) a true proposition of this science. All other statements can only be 'true' in the so-called popular sense; they are true only provisionally, as long as they have not yet been submitted to a scientific test. Only the latter will yield the definitive decision. If I say: 'He is a man of good character', this judgement will *never* become a scientific one because the concepts it presupposes are too complicated. It will therefore always remain a 'popular' one and hence only 'popularly true'. The same holds for the simpler statement 'My pipe is broken'. We continually operate with such

---

5 The copy retained by Dingler contains the following additional paragraph with an indication that it was crossed out.

judgements in practical life, but they are not, and never will be, scientific synthetic judgements, and they cannot therefore provide norms for the latter.

III/5 [ix/6]   DINGLER to FREGE   27.6.1917

Augsburg
27 June 1917

My dear Professor,

I do not know whether to conclude from your silence following my last letter that there was something in it that displeased you. We already knew in advance that our opinions would not be exactly the same. It therefore seemed all the more valuable to try to come to a mutual understanding, as you put it so well in your first letter, for all these things refer in the end to circumstances which are objective and can be seen objectively, so one only needs to establish a common ground. I freely admit that up to now I have not discussed your publications very much in my letters, because, frankly, I regarded the difference between us as so great that there was little to be done about it. But upon closer reflection I must confess that I have been strongly influenced by a more extraneous reason, which is that your writings have always been difficult for me to get access to. I bought a copy of your *Conceptual Notation* in 1905, but because of the way you presented your views I was unable to gain the degree of enlightenment about them I had hoped for. Unfortunately, Munich University Library does not have your main work,[1] and the Public Library lends books for such short periods that is is practically of no use. If I am to establish a real relationship with a book like yours, I must have it lying around for about a year. Of your essays, I have read the ones in the *Annual Reports of the German Mathematical Society* with great interest;[2] the others I do not know. But I do not always have access even to this journal. Our rather brief correspondence has now given me a strong desire to concern myself more closely with your views, which are now of great interest to me and which I am now perhaps in a better position to appreciate, but I have no possibility of doing this in my present isolation. Could you not send me a collection of your offprints? They will certainly gather less dust with me than with many other mathematicians who are less interested in these logical questions. I consider it too immodest to ask you for a copy of your main work, in spite of all the justifications inherent in such a request. Once I have a more precise knowledge of your entire

---

III/5. 1 GGA I–II.
2 Frege published GLG I–III3, as well as items ⟨24⟩, ⟨25⟩ and ⟨26⟩ which are part of his controversy with Thomae, in *Jahresberichte der Deutschen Mathematiker-Vereinigung*.

position, I shall be better placed to do justice to it than I am now, though I already have the best will, and I think that we shall then be able to carry our conversation by mail a little further, to the satisfaction of both parties and in pursuit of our common interest: truth.

Yours sincerely,
H. Dingler

III/6 [ix/7]   FREGE to DINGLER   4.7.1917

Brunshaupten on the Baltic
4 July 1917

My dear Captain,

Please excuse me for not having answered your letter. In any case, the reason was not that something in it had displeased me. So many different things have lately been going through my head that I quite forgot that I still owed you a reply. I also cannot find your letter here; I suppose it has been left behind in Jena. My *Conceptual Notation* is now already somewhat outdated and no longer corresponds entirely to my position. I am thinking of sending you some offprints. Unfortunately I can no longer do this for all of my essays. I will not send you any that appeared in the *Annual Report* since you probably know them.[1] The ones I can still find here are the following: in vol. XII: 'On the Foundations of Geometry';[2] the same in vol. II;[3] in vol. 15: 'On the Foundations of Geometry' I and II[4] and 'Reply to Mr Thomae's Holiday Chat';[5] in vol. 17: 'The Impossibility of Thomae's Formal Arithmetic'.[6] I am thinking of sending you: *Revue de métaphysique et de morale* (1895, No. 1);[7] 'On the Numbers of Mr H. Schubert';[8] Review of H. Cohen, *The Principle of the Infinitesimal Method and its History*;[9] 'On Formal Theories of Arithmetic';[10] and 'Applications of the Conceptual Notation'.[11] This lecture still represents the position of my *Conceptual Notation*. Instead of '≡' I would now write '='; for I see that the equals sign is used in mathematics as a sign of identity. In geometry, too, the sign '=' can at least be understood in the same way if '*AB*' is taken to mean not the

---

III/6. 1 Cf. III/5, note 2.
2 $= \langle 18 \rangle$.
3 $= \langle 19 \rangle$.
4 $= \langle 21 \rangle$ and $\langle 22 \rangle$.
5 $= \langle 24 \rangle$.
6 $= \langle 25 \rangle$.
7 This journal contains Frege's essay 'Le nombre entier' $\langle 12 \rangle$.
8 $= \langle 16 \rangle$.
9 $= \langle 5 \rangle$.
10 $= \langle 6 \rangle$.
11 $= \langle 2 \rangle$.

length but the measure of the length, or the number we get when we measure the length. I later use another sign in place of

$$\frac{\gamma}{\bar{\beta}} f(x_\nu, y_\beta)$$

but I find that this sign is perhaps preferable after all.[12] At the time I did not yet make the distinction between 'mean' and 'express' which I observed strictly later on. I therefore use the word 'means' several times in my lecture where I would now say 'expresses'. I think I first made the distinction in my essay. 'On Sense and Meaning' (in *Zeitschrift für Philosophie und philosophische Kritik*, vol. 100),[13] of which I no longer have an offprint. Perhaps you will also be interested in 'Function and Concept' (Jena, Hermann Pohle, 1891).[14] I only have one copy of it left, which I would like to keep. Further, I am thinking of sending you: 'What is a Function?'[15] and 'On the Conceptual Notation of Mr Peano and My Own'.[16] You will find references to many of my writings in Bertrand Russell, *Principles of Mathematics*.[17]

In the hope that this parcel will have something to offer you, I remain,

Yours sincerely,
G. Frege

III/7 [ix/8]   DINGLER to FREGE   10.7.1917

Augsburg
10 July 1917

My dear Councillor,

Please accept my sincere thanks for the very interesting material you sent me. I only regret that the essay from the *Revue de métaphysique et de morale* could not be fitted in.[1] I had already asked my publisher to send you a copy of my work on *The Principle of Logical Independence.**[2] Unfortunately I have only one copy left of my *Philosophy of Nature* in

---

* and I will send you something or other in the near future. Since I live provisionally out of boxes, I will first have to dig it out.

---

12 This formula, which occurs in ⟨2⟩, means that y is part of series f which begins with x. The 'later' usage mentioned here is probably that of GGA I, sects 45ff.

13 = ⟨9⟩, most commonly called 'On Sense and Reference'.

14 = ⟨7⟩.

15 = ⟨20⟩.

16 = ⟨15⟩.

17 Cambridge 1903; 2nd ed. London 1937.

III/7. 1 = ⟨12⟩.

2 *Das Prinzip der logischen Unabhängigkeit in der Mathematik, zugleich als Einführung in die Axiomatik* (Munich 1915).

which the title page and preface are missing because of a bookbinder's oversight.[3] But since the contents are complete, I will send it to you. At the moment I am reading with great interest your brilliant discussions in the essays you sent me. In case you have *spare* copies of the essays from the *Annual Report*, I should also be very grateful for these.[4]

Yours sincerely,
[H. Dingler]

III/8 [ix/9]   FREGE to DINGLER   21.7.1917

Brunshaupten on the Baltic
21 July 1917

My dear Doctor,

Many thanks for sending me your book on *The Principle of Logical Independence* and for your letter of the 10th.

At the same time I am sending you some more of my articles in the hope that they will be welcome.

Yours sincerely,
G. Frege

III/9 [ix/10]   FREGE to DINGLER   1.8.1917

My dear Captain,

Many thanks for your parcel. I have been looking into the *Foundations of the Philosophy of Nature*. Unfortunately it is slow going since there is much to distract me in various directions. Even though your way of expressing yourself and your use of individual words diverges from mine, it still seems to me that in matters of substance one can recognize some points of contact between our lines of thought.

Yours sincerely,
G. Frege

3 *Die Grundlagen der Naturphilosophie* (Leipzig 1913; reprinted Darmstadt 1967).
4 Cf. III/6, note 2.

III/10 [ix/15]   FREGE to DINGLER   17.11.1918

Bad Kleinen, Mecklenburg
17 November 1918

My dear Doctor,

Thank you very much for the kind lines and friendly wishes you sent me for my birthday. Yes, the search for truth seems to me at the very heart of my profession.

In these difficult times I seek consolation in scientific work. I am trying to bring in the harvest of my life so it will not be lost. I have written an essay for the *Beiträge zur Philosophie des deutschen Idealismus* which, I think, will appear shortly,[1] and a sequel to it which will perhaps be printed a year from now.[2]

I am also concerned with matters that fall into the area of politics and political economy, as for example with proposals for an electoral law. But here I have the depressing feeling that my work will be in vain. Nobody has time for this now. I have sent typewritten copies to representatives and other gentlemen, but have received only one reply – an affirmative one – from the président of the Farmers' Union, Dr Boehme. But there is no prospect that these proposals can ever be realized. Here I have of course stepped onto ground outside the field of my usual endeavours.

With all good wishes, I remain,

Yours sincerely,
G. Frege

III/10. 1 = ⟨27⟩.
        2 = ⟨28⟩.

IV FREGE–HILBERT

Editor's Introduction

David Hilbert (1862–1943) accepted a professorship in mathematics at Göttingen in 1895.[1] In the same year he met Frege at the Convention of German Scientists and Doctors in Lübeck and, following a lecture by Frege, had a conversation with him which is continued in the first two letters. In the winter semester 1898–9 Hilbert lectured on 'The Elements of Euclidean Geometry'. This lecture became the basis for his work *Grundlagen der Geometrie*, which appeared in 1899 as part of a *Festschrift* on the occasion of the unveiling of the Gauss–Weber monument. Frege may have had advance knowledge of Hilbert's views through an edited transcript of Hilbert's lectures prepared by H. von Schaper. This transcript reached Frege through Heinrich Liebmann (the son of Otto Liebmann, Frege's colleague at Jena), who was at the time a lecturer in mathematics at Göttingen and who had heard Hilbert's lectures. However, it is not certain that the transcript reached Frege prior to the publication of the work. Frege's objections to Hilbert's use of certain terms led to a more extended controversy between them which is contained in letters 3 to 8. Letter 9, which follows the others after an interval of three years, adds little to the controversy.

The Frege–Hilbert controversy has been much studied and discussed by writers on the foundations of mathematics. Although Frege's logical objections were well taken, and although a correct understanding of the axiomatic method must begin with Frege, the dominant view, especially among mathematicians, is still the view expressed recently by H. Scholz: '... no one doubts nowadays that while Frege himself created much that was radically new on the basis of the classical conception of science, he was no longer able to grasp Hilbert's radical transformation of this conception of science, with the result that his critical remarks, though very acute in themselves and still worth reading today, must nevertheless be regarded as essentially beside the point.'[2] While H. Freudenthal has tried to give a more balanced historical account, his study points nevertheless in the same direction.[3] H. G. Steiner does more justice to Frege's arguments, but his verdict, too, is largely determined by the traditional interpretation of Hilbert's school.[4] A contrary interpretation has been proposed by F. Kambartel, who argues that Frege's

1 On Hilbert's life and work, cf. the biography by O. Blumenthal in D. Hilbert, *Gesammelte Abhandlungen* III (1935, reprinted New York 1965), p. 402, and further C. Reid, *Hilbert* (Berlin etc. 1970).

2 H. Scholz, *Mathesis Universalis*, in *Abhandlungen zur Philosophie als strenge Wissenschaft*, H. Hermes et al., eds (Basel and Stuttgart 1961; 2nd ed. Darmstadt 1969), p. 222.

3 H. Freudenthal, *Zur Geschichte der Grundlagen der Geometrie. Zugleich eine Besprechung der 8. Aufl. von Hilberts 'Grundlagen der Geometrie'*, Nieuw Archief voor Wiskunde 5/6 (1957–8), pp. 105–42.

4 H. G. Steiner, 'Frege und die Grundlagen der Geometrie', in *Mathematik an Schule und Universität, H. Behnke zum 65. Geburtstag gewidmet*, K.-P. Grotemeyer et al., eds (Göttingen 1964), pp. 175–86 and 293–305.

objections can be construed almost in their entirety as a well-founded critique of Hilbert's ideas.[5]

In the fifth letter, Frege proposed to Hilbert that their correspondence be published, but Hilbert did not take up the suggestion. Frege later had the following comment on this:

> Mr Hilbert's *Festschrift* on the foundations of geometry prompted me to explain my divergent views to the author in writing; and this gave rise to a correspondence which, unfortunately, was soon broken off. Thinking that the questions it dealt with might be of general interest, I thought of a subsequent publication. However, Mr Hilbert is reluctant to give his consent to it since his own views have changed in the meantime. I regret this because reading the correspondence would have been the most convenient way of introducing someone into the state of the question, and it would have saved me the trouble of reformulating it . . .[6]

The last sentence makes it clear that Frege decided to continue the controversy in the form of an essay which appeared in 1903 under the title 'Über die Grundlagen der Geometrie' in two parts.[7] Hilbert himself did not react to it. A. Korselt undertook the defence of Hilbert's position,[8] and Frege replied to Korselt in a new series of articles entitled 'Über die Grundlagen der Geometrie'.[9]

IV/1 [xv/1]   FREGE to HILBERT   1.10.1895

Jena
1 October 1895

Dear Colleague,

If I remember right, you told me in Lübeck that you were trying to decrease rather than to increase formalization in mathematics.[1] Since we were interrupted in our conversation, I should like to explain my view to you

---

5 F. Kambartel, *Erfahrung und Struktur: Bausteine zu einer Kritik des Empirismus und Formalismus* (Frankfurt 1968), pp. 155ff.

6 (⟨18⟩), p. 319.

7 (⟨18⟩) and ⟨19⟩.

8 A. Korselt, 'Über die Grundlagen der Geometrie', *Jahresberichte der Deutschen Mathematiker-Vereinigung* 12 (1903), pp. 402–07.

9 ⟨21⟩ to ⟨23⟩.

IV/1. 1 Frege and Hilbert met at the 67th Convention of German Scientists and Doctors which was held in Lübeck from 16 to 20 September 1895. On 17 September Frege gave a lecture, in the Section on Mathematics and Astronomy, which became the basis of the lecture delivered to the Royal Saxon Society for the Sciences in Leipzig on 6 July 1896 and published in 1897 as 'Über die Begriffsschrift des Herrn Peano und meine eigene' ⟨15⟩. Frege's general remarks, both in Lübeck and in the published lecture, on the advantages of a conceptual notation over 'word languages' were presumably the occasion for Hilbert's remarks on 'formalization in mathematics'.

in writing. What is basically at issue is not, I believe, the contrast between spoken words and written symbols, but whether one ought to use theorems and methods with a wider or with a narrower scope. This contrast seems to coincide with the former only if there is as yet no adequate symbolism for the methods you rightly prefer: those with wide scope. But where a line of thought can be perfectly expressed in symbols, it will appear briefer and more perspicuous in this form than in words. Here I am presupposing that it is really the same line of thought, and that one is not following an entirely different method. Only then can the comparison be made. The advantages of perspicuity and precision are so great that many investigations could not even have been made without a mathematical sign language. Now it may well happen that, as science makes further advances, the same results can be reached more easily and more perfectly in different ways, without or with only little use of symbols. But if the sign language has been perfected to the point where it can express the new line of thought, the latter will appear more perspicuously in symbols than in words.

Further, the use of symbols must not be equated with a thoughtless, mechanical procedure, although the danger of lapsing into a mere mechanism of formulas is more immediate here than with the use of words. One can think also in symbols. A mere mechanical operation with formulas is dangerous (1) for the truth of the results and (2) for the fruitfulness of the science. The first danger can probably be avoided almost entirely by making the system of signs logically perfect. As far as the second danger is concerned, science would come to a standstill if the mechanism of formulas were to become so rampant as to stifle all thought. Yet I would not want to regard such a mechanism as completely useless or harmful. On the contrary, I believe that it is necessary. The natural course of events seems to be as follows: what was originally saturated with thought hardens in time into a mechanism which partly relieves the scientist from having to think. Similarly, in playing music, a series of processes which were originally conscious must have become unconscious and mechanical so that the artist, unburdened of these things, can put his heart into the playing. I should like to compare this to the process of lignification. Where a tree lives and grows it must be soft and succulent. But if what was succulent did not in time turn into wood, the tree could not reach a significant height. On the other hand, when all that was green has turned into wood, the tree ceases to grow.

The natural way in which one arrives at a symbolism seems to me to be this: in conducting an investigation in words, one feels the broad, imperspicuous and imprecise character of word language to be an obstacle, and to remedy this, one creates a sign language in which the investigation can be conducted in a more perspicuous way and with more precision. Thus the need comes first and then the satisfaction. The contrary approach, that of first creating a symbolism and then looking for an application for it, would seem to be less beneficial. Perhaps the symbolism of Boole, Schröder, and Peano has taken this path.

In the hope that I have not bored you with these explanations, I remain,

Yours sincerely,

Dr G. Frege

IV/2 [xv/2]   HILBERT to FREGE   4.10.1895

Göttingen

4 October 1895

Dear Colleague,

Your valuable letter was of extraordinary interest to me; I regret all the more that I only spoke to you so briefly in Lübeck. I hope there will be more of an opportunity some other time.

I intend to bring up your letter for discussion in our mathematical society at the beginning of the semester. I believe that your view of the nature and purpose of symbolism in mathematics is exactly right. I agree especially that the symbolism must come later and in response to a need, from which it follows, of course, that whoever wants to create or develop a symbolism must first study those needs. With best regards,

Yours sincerely,

Hilbert

IV/3 [xv/3]   FREGE to HILBERT   27.12.1899

Jena

27 December 1899

Dear Colleague,

I was interested to get to know your *Festschrift* on the foundations of geometry,[1] the more so as I myself had earlier been concerned with them, but without publishing anything. As was to be expected, there are many points of contact between my earlier unrealized attempts and your account, but also many divergencies. In particular, I thought I could make do with fewer primitive terms. When I spoke with my colleagues Thomae and Gutzmer about your work,[2] it turned out that we were not always clear about your real meaning, and this prompts me to put my doubts before you

---

IV/3. 1 Frege is referring to the first edition of Hilbert's *Grundlagen der Geometrie* [E.T. *Foundations of Geometry*, Chicago, 1901] which appeared as part I (pp. 1–92) of a *Festschrift zur Feier der Enthüllung des Gauss-Weber-Denkmals in Göttingen* (Leipzig 1899).

2 C. F. A. Gutzmer and J. Thomae, both professors of mathematics at the University of Jena at the time, took a 'formalist' view of the foundations of arithmetic, a view Frege criticized in great detail (cf. GGA II, sects 86ff, ⟨24⟩, ⟨25⟩ and ⟨26⟩.

in the hope that you will not be unwilling to enlighten me about them. Let me start with something Thomae said about your explanation in sect. 3.[3] His words were roughly: 'That is not a definition; for it does not give a characteristic mark by which one could recognize whether the relation Between obtains'. I, too, cannot take it for a definition, but you also do not call it a definition but an explanation. You evidently use the two expressions 'explanation' and 'definition' to designate different things, but the difference is not clear to us. The explanations of sect. 4 seem to be of exactly the same kind as your definitions: we are told, e.g., what the words 'lie on line a on the same side as point 0' are supposed to mean, just as we are told in the following definition what the word 'line section' is supposed to mean.[4] The explanations of sects 1 and 3 are apparently of a very different kind, for here the meanings of the words 'point', 'line', 'between' are not given, but are assumed to be known in advance. At least it seems so. But it is also left unclear what you call a point. One first thinks of points in the sense of Euclidean geometry, a thought reinforced by the proposition that the axioms express fundamental facts of our intuition.[5] But afterwards (p. 20) you think of a pair of numbers as a point.[6] I have my doubts about the proposition that a precise and complete description of relations is given by the axioms of geometry (sect. 1)[7] and that the concept 'between' is defined by axioms (sect. 3). Here the axioms are made to carry a burden that belongs to definitions. To me this seems to obliterate the dividing line between definitions and axioms in a dubious manner, and beside the old meaning of the word 'axiom', which comes out in the proposition that the axioms express fundamental facts of intuition, there emerges another

---

3 In the following passage, Frege is referring mainly to two propositions which Hilbert places before the group of axioms of order II/1–II/4: (1) 'The axioms of this group define the concept "between" and make it possible to *order* the points on a line, in a plane, and in a volume on the basis of this concept.' (2) 'Explanation. The points on a line stand in certain relations to one another which can be described by making use of the word "between".'

4 In the 'explanations' of sect. 4 Hilbert introduces certain terms by stating that a line is divided by a point, a plane by a line, and a volume by a plane into two 'sides'. On the other hand, in the same section and for no apparent reason, Hilbert refers to his introduction of the terms 'line section', 'polygon', etc. as 'definitions'.

5 Cf. *Grundlagen der Geometrie*, sect. 1: 'The axioms of geometry are divided into five groups; every one of these groups expresses certain closely related fundamental facts of our intuition.'

6 Frege is here referring to Hilbert's consistency proof for his axiomatic system by means of an analytical model, a proof given in sect. 9 of *Grundlagen der Geometrie*. Hilbert there writes: 'Let us think of a pair of numbers (x, y) of domain Ω as a point …'

7 Hilbert writes: 'We think of points, lines, planes as standing in certain mutual relations and designate these relations by words like "lie", "between", "parallel", "congruent", "constant"; the precise and complete description of these relations is given by the *axioms of geometry*'.

meaning but one which I can no longer quite grasp. There is already widespread confusion with regard to definitions in mathematics, and some seem to act according to the rule:

> If you can't prove a proposition,
> Then treat it as a definition.

In view of this, it does not seem to me a good thing to add to the confusion by also using the word 'axiom' in a fluctuating sense and similar to the word 'definition'. I think it is about time that we came to an understanding about what a definition is supposed to be and do, and accordingly about the principles to be followed in defining a term (see my *Basic Laws of Arithmetic*, vol. I, sect. 33). It seems to me that complete anarchy and subjective caprice now prevail. Allow me to explain to you some of the thoughts I have had on this.

I should like to divide up the totality of mathematical propositions into definitions and all the remaining propositions (axioms, fundamental laws, theorems). Every definition contains a sign (an expression, a word) which had no meaning before and which is first given a meaning by the definition. Once this has been done, the definition can be turned into a self-evident proposition which can be used like an axiom. But we must not lose sight of the fact that a definition does not assert anything but lays down something. Thus we must never present as a definition something that is in need of proof or of some other confirmation of its truth. I use the equals sign as the sign of identity. Now suppose we know the meanings of the plus sign and the signs for three and one but not the meaning of the sign for four, then we can give the sign for four a meaning through the equation '$3 + 1 = 4$'. Once this has been done, this equation is self-evidently true and no longer in need of proof. But if one wanted to discharge the burden of proof by giving a definition, this would be logical sleight-of-hand. It is very essential for the strictness of mathematical investigations that the difference between definitions and all other propositions be observed in all strictness. The other propositions (axioms, fundamental laws, theorems) must not contain a word or sign whose sense and meaning, or whose contribution to the expression of a thought, was not already completely laid down, so that there is no doubt about the sense of the proposition and the thought it expresses. The only question can be whether this thought is true and what its truth rests on. Thus axioms and theorems can never try to lay down the meaning of a sign or word that occurs in them, but it must already be laid down. One can also recognize a third kind of proposition, elucidatory propositions, but I would not want to count them as part of mathematics itself but refer them to the antechamber, the propaedeutics. They are similar to definitions in that they too are concerned with laying down the meaning of a sign (or word). Thus they too contain something whose meaning cannot be assumed to be known in advance, at least not completely and beyond doubt, perhaps because they are used in the language of life in a fluctuating way or in many senses. If in such a case the meaning to be assigned is logically simple, then one cannot give a proper definition but must confine oneself to warding off the unwanted meanings among those that occur in linguistic usage and to pointing to the wanted one, and here one must of course always rely on being met half-way by an intelligent guess. Unlike definitions, such elucidatory propositions cannot be used in proofs because they lack the necessary precision, which is why I should like to refer them to the antechamber, as I said above.[8] I call axioms propositions that are true but are not proved because our knowledge of them flows from a source very different from the logical source, a source which might be called spatial intuition. From the truth of the axioms it follows that they do not contradict one another. There is therefore no need for a further proof. The definitions, too, must not contradict one another. If they do, they are faulty. The principles of definition must be such that if we follow them no contradiction can appear. If I was to set up your axiom II/1 as such,[9] I would presuppose a complete and unequivocal knowledge of the meanings of the expressions 'something is a point on a line' and '*B* lies between *A* and *C*', and in the latter case also a general knowledge of what is to be understood by the letters. Then the axiom cannot serve, e.g., to give a more precise explanation of the word 'between', and it is obviously impossible to give this word another meaning afterwards, as you seem to want to do on p. 20.[10] If this meaning is different from the meaning of the word 'between' in sect. 3, then you have a highly suspect ambiguity. This seems to leave one with no alternative but to assume that the word 'between' has as yet no meaning at all in II/1. But then II/1 cannot be true, and hence it cannot be an axiom in my sense of the word which is, I think, the generally accepted sense. But if that word had as yet no sense at all in sect. 3, as is probable, then the proposition II/1 also has no sense, does not express a thought, and hence does not express a fundamental fact of our intuition either. But what, then, is its purpose? Is it supposed to lay down the meaning of 'between' like a definition? But then this cannot be done again later on. In sect. 6 you say:[11] 'The axioms of this group define the concept of congruence or movement.' Why, then, are they not called definitions? What difference, if any, is there then between definitions and axioms? Indeed, the latter do not satisfy what is required of a definition, if only because there is more than one of them, and furthermore, because they

---

8 On Frege's distinction between elucidations and definitions in the narrower sense, cf. pw p. 207, and further, with reference to Hilbert, GLG III1, p. 301 [E.T. p. 58].
9 In the first edition of *Grundlagen der Geometrie*, axiom II/1 reads: 'If *A*, *B*, and *C* are points on a line and *B* lies between *A* and *C*, then *B* also lies between *C* and *A*.'
10 What is in question here is the introduction of order into Hilbert's model of his axiomatic system.
11 = sect. 5 in the 9th edition of *Grundlagen der Geometrie*.

contain expressions ('on a given side of line a'')[12] whose meanings seem not yet to have been laid down. I am well aware that to prove the mutual independence of the axioms you must adopt a higher standpoint from which Euclidean geometry appears as a special case of a more general theory; but for the reasons I have given it seems to me that the path you are taking is not open to you without further ado.

I would not regard your work as a valuable one if I did not believe I could see roughly how such objections could be rendered harmless; but this will not be possible without considerable reshaping. The first thing that seems to me necessary is to come to an understanding about the expressions 'explanation', 'definition', and 'axiom', where you strongly diverge from what is familiar to me as well as from what is customary, which makes it difficult for me to keep these expressions apart in your account and to see the logical structure with full clarity. In spite of these doubts, I am very much interested in your work, and it would give me great pleasure if you were to write me a letter explaining your position with respect to my doubts.

Please forgive these remarks and be assured that I have not made them to burden you with my doubts but because I believe that other readers of your work will have had these or similar doubts and that it would be desirable to dispel or to disarm them.

Yours sincerely,
Dr G. Frege

IV/4 [xv/4]   HILBERT to FREGE   29.12.1899

[Excerpt by Frege]

... One more preliminary remark: if we want to understand each other, we must not forget that the intentions that guide the two of us differ in kind. It was of necessity that I had to set up my axiomatic system: I wanted to make it possible to understand those geometrical propositions that I regard as the most important results of geometrical inquiries: that the parallel axiom is not a consequence of the other axioms, and similarly Archimedes' axiom, etc. I wanted to answer the question whether it is possible to prove the proposition that in two identical rectangles with an identical base line the sides must also be identical, or whether as in Euclid this proposition is a new postulate.* I wanted to make it possible to understand and answer such

---

* This proposition is after all the foundation of all planimetry.

---

12 The expression occurs in Hilbert's axiom IV/1 (=III/1 in the 9th edition of *Grundlagen der Geometrie*).

questions as why the sum of the angles in a triangle is equal to two right angles and how this fact is connected with the parallel axiom. That my system of axioms allows one to answer such questions in a very definite manner, and that the answers to many of these questions are very surprising and even quite unexpected, is shown, I believe, by my *Festschrift* as well as by the writings of my students who have followed it up. Among these I will refer only to Mr Dehn's dissertation which is to be reprinted shortly in *Mathematische Annalen*.[1] So much for my main intention. Of course, I also believe I have set up a system of geometry which satisfies the strictest demands of logic, and this brings me to the proper answer to your letter.

You say my explanation in sect. 3 is not a definition of the concept 'between', since it fails to give its characteristic marks. But these characteristic marks are given explicitly in axioms II/1 to II/5. However, if one wants to use the word 'definition' precisely in the customary sense, one will have to say:

'Between' is a relation which holds for the points on a line and which has the following characteristic marks: II/1 ... II/5.

You say further: 'The explanations in sect. 1 are apparently of a very different kind, for here the meanings of the words "point", "line", ... are not given, but are assumed to be known in advance.' This is apparently where the cardinal point of the misunderstanding lies. I do not want to assume anything as known in advance; I regard my explanation in sect. 1 as the definition of the concepts point, line, plane – if one adds again all the axioms of groups I to V as characteristic marks. If one is looking for other definitions of a 'point', e.g., through paraphrase in terms of extensionless, etc., then I must indeed oppose such attempts in the most decisive way; one is looking for something one can never find because there is nothing there; and everything gets lost and becomes vague and tangled and degenerates into a game of hide-and-seek. If you prefer to call my axioms characteristic marks of the concepts which are given and hence contained in the 'explanations', I would have no objection to it at all, except perhaps that it conflicts with the customary practice of mathematicians and physicists; and I must of course be free to do as I please in giving characteristic marks. For as soon as I have laid down an axiom, it exists and is 'true'; and this brings me now to a further important point in your letter. You write: 'I call axioms propositions ... From the truth of the axioms it follows that they do not contradict one another.' I found it very interesting to read this very sentence in your letter, for as long as I have been thinking, writing and lecturing on these things, I have been saying the exact reverse: if the arbitrarily given axioms do not contradict one another with all their consequences, then they are true and the things defined by the axioms exist. This is for me the criterion of truth

IV/4. 1 M. Dehn, 'Die Legendreschen Sätze über die Winkelsumme in Dreieck', *Mathematische Annalen* 53 (1900), pp. 404–39.

and existence. The proposition 'Every equation has a root' is true, and the existence of a root is proven, as soon as the axiom 'Every equation has a root' can be added to the other arithmetical axioms, without raising the possibility of contradiction, no matter what conclusions are drawn. This conception is indeed the key to an understanding not just of my *Festschrift* but also for example of the lecture I recently delivered in Munich on the axioms of arithmetic,[2] where I prove or at least indicate how one can prove that the system of all ordinary real numbers *exists*, whereas the system of all Cantorian cardinal numbers or of all alephs does *not* exist – as Cantor himself asserts in a similar sense and only in somewhat different words. Thus, to restate the main point: the renaming of 'axioms' as 'characteristic marks' is surely an extraneous matter as well as a matter of taste – and it is in any case easily done. On the other hand, to try to give a definition of a point in three lines is to my mind an impossibility, for only the whole structure of axioms yields a complete definition. Every axiom contributes something to the definition, and hence every new axiom changes the concept. A 'point' in Euclidean, non-Euclidean, Archimedean and non-Archimedean geometry is something different in each case. After a concept has been fixed completely and unequivocally, it is on my view completely illicit and illogical to add an axiom – a mistake made very frequently, especially by physicists. By setting up one new axiom after another in the course of their investigations, without confronting them with the assumptions they made earlier, and without showing that they do not contradict a fact that follows from the axioms they set up earlier, physicists often allow sheer nonsense to appear in their theoretical investigations. One of the main sources of mistakes and misunderstandings in modern physical investigations is precisely the procedure of setting up an axiom, appealing to its truth (?), and inferring from this that it is compatible with the defined concepts. One of the main purposes of my *Festschrift* was to avoid this mistake.

There is only one more objection I must touch on. You say that my concepts, e.g. 'point', 'between', are not unequivocally fixed; e.g. 'between' is understood differently on p. 20, and a point is there a pair of numbers. But it is surely obvious that every theory is only a scaffolding or schema of concepts together with their necessary relations to one another, and that the basic elements can be thought of in any way one likes. If in speaking of my points I think of some system of things, e.g. the system: love, law, chimney-sweep ... and then assume all my axioms as relations between these things, then my propositions, e.g. Pythagoras' theorem, are also valid for these things. In other words: any theory can always be applied to infinitely many

---

2 'Über den Zahlbegriff', published in *Jahresberichte der Deutschen Mathematiker-Vereinigung* 8 (1900), pp. 180–4, and reprinted as Appendix VI to the 7th edition of Hilbert's *Grundlagen der Geometrie*.

systems of basic elements. One only needs to apply a reversible one-one transformation and lay it down that the axioms shall be correspondingly the same for the transformed things. This circumstance is in fact frequently made use of, e.g. in the principle of duality, etc., and I have made use of it in my independence proofs. All the statements of the theory of electricity are of course also valid for any other system of things which is substituted for the concepts magnetism, electricity . . ., provided only that the requisite axioms are satisfied. But the circumstance I mentioned can never be a defect in a theory,* and it is in any case unavoidable. However, to my mind, the application of a theory to the world of appearances always requires a certain measure of good will and tactfulness; e.g., that we substitute the smallest possible bodies for points and the longest possible ones, e.g. light-rays, for lines. We also must not be too exact in testing the propositions, for these are only theoretical propositions. At the same time, the further a theory has been developed and the more finely articulated its structure, the more obvious the kind of application it has to the world of appearances, and it takes a very large amount of ill will to want to apply the more subtle propositions of plane geometry or of Maxwell's theory of electricity to other appearances than the ones for which they were meant . . .

[Draft or Excerpt by Hilbert]

Göttingen
29 December 1899

Reply to letter from Frege, Jena.

My intention in composing the *Festschrift* was: to make it possible to understand the most beautiful and important propositions of geometry (unprovability of the parallel axiom, of Archimedes' axiom, provability of the Killing–Stolz axiom, etc.), so as to make it possible to give definite answers (some of which turn out very unexpected).

Instead of 'axioms' you can say 'characteristic marks' if you like. But if one is looking for another definition of, e.g., 'points', perhaps through paraphrase in terms of extensionless . . ., then I reject such attempts as fruitless, illogical and futile. One is looking for something where there is nothing. The whole investigation becomes vague and tangled and degenerates into a game of hide-and-seek. Definitions (i.e., explanations, definitions, axioms) must contain everything required for the construction of a theory, and they must contain only this. My division into explanations, definitions and axioms, which together make up definitions in your sense, does contain something arbitrary, but I believe that my arrangement is in general useful and perspicuous.

---

* It is rather a tremendous advantage.

I was very much interested in your sentence: 'From the truth of the axioms it follows that they do not contradict one another', because for as long as I have been thinking, writing, lecturing about these things, I have been saying the exact reverse: If the arbitrarily given axioms do not contradict one another, then they are true, and the things defined by the axioms exist. This for me is the criterion of truth and existence. The proposition 'Every equation has a root' is true, i.e., a proof of the existence of a root, if the proposition, when added as an axiom to the axioms of arithmetic, can never lead to a contradiction. E.g., the existence of real numbers and the non-existence of the system of all cardinal numbers. Therefore: the definition of the concept point is not complete till the structure of the system of axioms is complete. For every axiom contributes something to the definition, and hence every new axiom changes the concept. A point in Euclidean, non-Euclidean, Archimedean and non-Archimedean geometry is something different in each case. After a concept has been fixed completely and unequivocally, it is completely illicit and illogical to add an axiom – a mistake which is very often made by physicists. Because they lay down new axioms throughout and do not confront them with the earlier ones, they come out with sheer nonsense. It is precisely the procedure of laying down an axiom, appealing to its truth (?), and then inferring from this that it is compatible with the defined concepts that is the eternal source of errors and misunderstandings. This is precisely what I wanted to avoid in my *Festschrift*.

There remains one objection to be touched on: you say that my concepts, e.g., 'point', 'between', are not unequivocally fixed. 'Between' is understood differently on p. 20, and a point there is a pair of points. But it is surely obvious that every theory is only a scaffolding (schema) of concepts together with their necessary connections, and that the basic elements can be thought of in any way one likes. E.g., instead of points, think of a system of love, law, chimney-sweep ... which satisfies all axioms; then Pythagoras' theorem also applies to these things. Any theory can always be applied to infinitely many systems of basic elements. For one only needs to apply a reversible one–one transformation and then lay it down that the axioms shall be correspondingly the same for the transformed things (as illustrated in the principle of duality and by my independence proofs). All statements of electrostatics hold of course also for any other system of things which is substituted for quantity of electricity ..., provided the requisite axioms are satisfied. Thus the circumstance I mentioned is never a defect (but rather a tremendous advantage) of a theory. The application of a theory to the world of appearances requires a measure of tact and good will. Points should be the smallest possible bodies, lengths the longest possible ones, e.g. rays. Also the testing of propositions should not be too meticulous. At the same time, the further a theory has been developed and the finer its ramifications, the more obvious the kind of application it has to the world of appearances. It takes a large measure of ill will to want to apply the subtler propositions of plane geometry or the propositions of Maxwell's theory of electricity to other appearances than the ones for which they were meant.

Signed Hilbert

IV/5 [xv/5]   FREGE to HILBERT   6.1.1900

Jena
6 January 1900

Dear Colleague,

Thank you very much for your long and interesting letter, which gave me much pleasure. I see that we have often thought about the same questions but have not always arrived at the same results. Our exchange of ideas promises to become all the more fruitful (at least for me), and I am glad that you would like me to reply.

Thank you very much for sending me your Munich lecture: I believe that from it I can see a little more clearly what your plan is. It seems to me that you want to detach geometry entirely from spatial intuition and to turn it into a purely logical science like arithmetic. The axioms which are usually taken to be guaranteed by spatial intuition and placed at the base of the whole structure are, if I understand you correctly, to be carried along in every theorem as its conditions – not of course in a fully articulated form, but included in the words 'point', 'line', etc. You want to prove the mutual independence and lack of contradiction of certain premises (axioms), as well as the unprovability of propositions from certain premises (axioms). From a general logical point of view the case is always the same: you want to show the lack of contradiction of certain determinations. 'D is not a consequence of A, B and C' says the same thing as 'The satisfaction of A, B and C does not contradict the non-satisfaction of D'. 'A, B and C are independent of one another' means 'C is not a consequence of A and B; B is not a consequence of A and C; A is not a consequence of B and C'. After reducing everything to the same schema in this way, we must ask, What means have we of demonstrating that certain properties, requirements (or whatever else one wants to call them) do not contradict one another? The only means I know is this: to point to an object that has all those properties, to give a case where all those requirements are satisfied. It does not seem possible to demonstrate the lack of contradiction in any other way. If you are merely concerned to demonstrate the mutual independence of axioms, you will have to show that the non-satisfaction of one of these axioms does not contradict the satisfaction of the others. (I am here adopting your way of using the word 'axiom'.) But it will be impossible to give such an example in the domain of elementary Euclidean geometry because all the axioms are true in this domain. By placing yourself in a higher position from which Euclidean geometry appears as a special case of a more comprehensive theoretical structure, you widen your view so as to include examples which make the mutual independence of those axioms evident. Although I am struck here by a doubt, I will not pursue it further here. The main point seems to me to be that you want to place Euclidean geometry under a higher point of view. And indeed, the mutual independence of the axioms, if it can be proved at all, can only be proved in this way. Such an undertaking seems to me to be of the greatest scientific interest if it refers to the axioms in the old traditional sense of elementary Euclidean geometry. If such an undertaking extends to a system of propositions which are arbitrarily set up, it should *in general* be of far less scientific importance. Whether it is possible to prove the mutual independence of the axioms of Euclidean geometry in this way, I dare not decide, because of the doubt I indicated above. But even if you do not succeed, your idea of regarding Euclidean geometry as a special case of a more comprehensive theory is nevertheless a valuable one.

I fully agree with you that the genetic method leaves something to be desired as far as logical certainty is concerned.[1] It lies in the nature of things that science has developed along this line; but this should not make us forget that we must always keep in view a logically perfect system as the goal towards which this development tends. You write: 'After a concept has been fixed completely and unequivocally, it is on my view completely illicit and illogical to add an axiom.' If I understand your view correctly here, I can only agree with joy, the more so as I believed I was almost alone in holding this view. Indeed, the defect of the genetic method lies precisely in this: that the concepts are not ready and are nevertheless used in this less than ready and hence not properly usable state, and that we never know whether a concept is finally ready. So it happens that after a proposition has been proved it becomes false again because of the continued development, for the thought contained in the proposition becomes a different one. Such changes are especially dangerous, for since the wording remains the same, one does not even become fully aware of the change.

I also agree with you that definitions of a 'point' which paraphrase it in terms of 'extensionless' are of little value; but I would not be reluctant to confess that a 'point' cannot properly be defined at all.

As long as we are dealing with generalities, we seem to be in satisfactory agreement. This is no longer so when we turn to the actual execution. I had envisaged beforehand the possibility of conceiving your axioms as component parts of your explanations, and yet I was surprised to learn that all the axioms of groups I to V are to be taken as supplements to the explanation in sect. 1. Accordingly, this explanation, with all that belongs to it, fills all of your chapter I and interlocks with many other explanations and

---

IV/5. 1 In his Munich lecture, Hilbert criticized the method of deriving the system of real numbers 'genetically' from the natural numbers through 'successive expansion' and instead advocated the 'axiomatic method' also for arithmetic, 'to give a definitive representation and to assure complete logical certainty of the content of our knowledge' (*Grundlagen der Geometrie*, 7th ed. (1930), pp. 241ff).

theorems contained in the chapter. I confess that this logical edifice strikes me as mysterious and extremely imperspicuous. In thinking about definition, I have been tightening my requirements more and more, to the point where I have moved so far from the opinions of most mathematicians that communication has become very difficult. What I object to will perhaps become clearest if we consider your explanation in sect. 3. For the sake of comparison let me adduce Gauss's definition of numerical congruence. If we know what a difference is and what it means for one number to 'go into' another, then this explanation gives us a complete mastery of the concept of numerical congruence, and we can decide at once, e.g., whether $2 \equiv 8 \pmod{3}$.[2] The matter would be quite different if the word 'congruent' was explained not only through something known but also through itself, by saying in your manner:

'Explanation. Whole numbers stand in certain relations to one another, which can be described by making use of the word "congruent".

'Axiom 1. Any number is congruent with itself according to any modulus.

'Axiom 2. If one number is congruent with a second and the second with a third according to the same modulus, then the first is also congruent with the third according to this modulus.' Etc.

Should we be able to gather from such a definition that $2 \equiv 8 \pmod{3}$? Hardly! The matter is even worse if we turn to your explanation of 'between'; for your axioms of order also contain the words 'point' and 'line' whose meanings are also unknown. Your system of definitions is like a system of equations with several unknowns, where there remains a doubt whether the equations are soluble and, especially, whether the unknown quantities are uniquely determined. If they were uniquely determined, it would be better to give the solutions, i.e., to explain each of the expressions 'point', 'line', 'between' individually through something that was already known. Given your definitions, I do not know how to decide the question whether my pocket watch is a point. The very first axiom deals with two points;[3] thus if I wanted to know whether it held for my watch, I should first have to know of some other object that it was a point. But even if I knew this, e.g., of my penholder, I still could not decide whether my watch and my penholder determined a line, because I would not know what a line was. The word 'determine' would then also create difficulties. But even if I understood the words 'point' and 'line' as in elementary geometry and I was given three points on a line, I still could not decide from your definition and the pertinent

---

2 '$x \equiv y \pmod{m}$' – or in words, x and y are congruent for modulus m – is true by definition if and only if m divides $x - y$ (or 'goes into' $x - y$).
3 Hilbert's axiom I/1 reads: 'Two different points A and B always determine a line a; we assume $AB = a$ or $BA = a$.'

axioms which of these points lay between the two others, nor would I know what investigations I should have to pursue for this purpose. To this must be added the following: According to I/7, there are at least two points on a line.[4] But what would you say about the following:

'Explanation. We imagine objects we call Gods.
'Axiom 1. All Gods are omnipotent.
'Axiom 2. All Gods are omnipresent.
'Axiom 3. There is at least one God.'

Here we must consider my distinction between first- and second-level concepts. I say 'my' distinction because as far as I know it has not been drawn sharply enough before me. In the words 'there is' we have a second-level concept, which must not be combined with the first-level concepts *omnipotent* and *omnipresent* as a characteristic mark of a first-level concept. (See my *Foundations of Arithmetic*, sect. 53, where I speak of 'order' instead of 'level', and my *Basic Laws of Arithmetic*, sects 21 and 22.) The characteristic marks you give in your axioms are apparently all higher than first-level; i.e., they do not answer to the question 'What properties must an object have in order to be a point (a line, plane, etc.)?', but they contain, e.g., second-level relations, e.g., between the concept *point* and the concept *line*. It seems to me that you really want to define second-level concepts but do not clearly distinguish them from first-level ones. The objectionable character of the definition of quantity given, e.g., by Stolz in his introduction to his *Lectures on General Arithmetic* flows apparently from the same source.[5]

---

4 The reference is to Hilbert's axiom I/7: 'On any line there are at least two points, in any plane at least three points which are not on a line, and in any volume at least four points which do not lie in a plane.'

5 Frege seems to be referring mainly to the following statements by Stolz:

The term 'quantity' (μέγεθος) occurs in Euclid's *Elements*, but he nowhere explains the concept. According to him we are to regard bounded geometrical forms — lines, angles, areas, bodies — as quantities, but also, it seems, the natural numbers. The following characteristic marks are common to all and make all calculation with them possible: those quantities that are of the same kind can be compared with one another, added, subtracted, and in general, any quantity can be divided into parts of the same kind as the whole. The geometry of the ancients contrasts quantities with geometrical relationships, even though it also attributes the first of the above characteristic marks to the latter and, as we shall see, could have added all the other characteristic marks as well. By ascending from geometrical quantities, natural numbers, and Euclidean relations to the next higher genus, we arrive at *absolute quantity in the narrower sense* ... Nothing prevents us from setting up even more general concepts. If we hold on only to the first characteristic mark in the above group, we get the widest concept, which H. Grassmann too takes as his starting-point. Let us give it the name 'quantity'; then the concept of a quantity will be such that any two of the things falling under it are defined as either the same or not the same. In other words, 'quantity is anything which is either to be equated with another thing as the same or differentiated from it as not the same'. All things which are comparable with *one*

Constant and characteristic here is the occurrence of the words 'of the same kind' with a completely blurred meaning. This can be avoided only by completely changing the way the question is put. And to repair the damage that I believe I find in your definitions one will have to proceed in a similar way; except that it will be much more difficult because what is in question is not just a single system but three systems (of points, lines and planes) with their manifold relations. Incidentally, what are you calling a system here? I believe it is the same thing which is elsewhere called a set or a class and which is best called the extension of a concept.

Our views are perhaps most sharply opposed with regard to your criterion of existence and truth. But perhaps I do not understand your meaning perfectly. To get clear about it, I present the following example:

Suppose we knew that the propositions

(1) A is an intelligent being
(2) A is omnipresent
(3) A is omnipotent

together with all their consequences did not contradict one another; could we infer from this that there was an omnipotent, omnipresent, intelligent being? This is not evident to me. The principle would read as follows:

If the propositions

'A has property Φ'
'A has property Ψ'
'A has property X'

together with all their consequences do not contradict one another in general (whatever A may be), then there is an object which has all the properties Φ, Ψ and X. This principle is not evident to me, and if it was true it would probably be useless. Is there some other means of demonstrating lack of contradiction besides pointing out an object that has all the properties? But if we are given such an object, then there is no need to demonstrate in a roundabout way that there is such an object by first demonstrating lack of contradiction.

If a general proposition contains a contradiction, then so does any particular proposition that is contained in it. Thus if the latter is free from contradiction, we can infer that the general proposition is free from contradiction, but not conversely. Suppose we have proved that in an

---

thing are called quantities *of the same kind* (or homogeneous quantities) and form a system of quantities.
(Cf. Otto Stolz, *Vorlesungen über allgemeine Arithmetik* I (Leipzig 1885), pp. 1ff. The quoted sentence is a citation from H. Grassmann, *Lehrbuch der Arithmetik* (Berlin 1861), p. 1.)

equilateral, rectangular triangle the square on the hypotenuse is twice as large as the square on one of the sides, which is easier than proving Pythagoras' theorem in general. We can now conclude further that the proposition does not contain a contradiction either within itself or in relation to the geometrical axioms. But can we now conclude further: therefore, Pythagoras' theorem is true? I cannot accept such a method of inference from lack of contradiction to truth. But this is probably not what you mean either. At any rate, a more precise formulation seems necessary.

It also seems to me that there is a logical danger in your speaking of, e.g., 'the parallel axiom', as if it was the same thing in every special geometry. Only the wording is the same; the thought content is different in every different geometry. It would not be correct to call the special case of Pythagoras' theorem mentioned above *the* theorem of Pythagoras; for after proving that special case, one still has not proved Pythagoras' theorem. Now given that the axioms in special geometries are all special cases of general axioms, one can conclude from lack of contradiction in a special geometry to lack of contradiction in the general case, but not to lack of contradiction in another special case.

I will reserve the right to reply to what you say about the applicability of a theory and the reversible one-one transformation.

In conclusion, I should like to propose that we envisage an eventual publication of our correspondence, in view of the great importance of the questions we have discussed for the whole of mathematics.

In returning your greetings and best wishes for the new century, I remain,

Yours sincerely,

G. Frege

IV/6 [xv/6]   Hilbert to Frege   15.1.1900

Göttingen
15 January

Dear Colleague,

As I am at the moment overburdened with all kinds of work, it is unfortunately impossible for me to reply in detail to your letter. Your discussion is of much interest and great value to me. It will at least prompt me to think more precisely and to formulate my thoughts more carefully.

With best regards,

Yours sincerely,
Hilbert

IV/7 [xv/7]   FREGE to HILBERT   16.9.1900

Jena

16 September 1900

Dear Colleague,

Many thanks for the two publications which you kindly sent me.[1] I have read your 'Mathematical Problems' with great interest, undiminished by several differences of opinion. However, the area of friction between our opinions is already large enough, so that it is probably better if I refrain from extending it for the time being. I will therefore allow myself only a few remarks which are connected with the questions dealt with in our correspondence.

On p. 12 I was struck by a sentence in which you say about the axioms that they contain a precise and complete description of those relations that obtain between the elementary concepts of a science.[2] I cannot reconcile this with what you write in your letter about the axioms, according to which they are to be regarded as component parts of the definitions of those elementary concepts. There can be talk about relations between concepts – e.g., the subordination of one concept to another – only after these concepts have been given sharp limits, but not while they are being defined.

I believe I can deduce, from some places in your lectures, that my arguments failed to convince you, which makes me all the more anxious to find out your counter-arguments. It seems to me that you believe yourself to be in possession of a principle for proving lack of contradiction which is essentially different from the one I formulated in my last letter and which, if I remember right, is the only one you apply in your *Foundations of Geometry*.[3] If you were right in this, it could be of immense importance, though I do not believe in it as yet, but suspect that such a principle can be

---

IV/7. 1 One of the two offprints may have been Hilbert's paper 'Über den Zahlbegriff', which was based on his Munich lecture (cf. IV/4, note 2). The other was an offprint of Hilbert's famous lecture, 'Mathematische Probleme', which was delivered to the International Congress of Mathematicians, Paris 1900, and first published in *Nachrichten der Königlichen Gesellschaft der Wissenschaften zu Göttingen, Mathematisch-physikalische Klasse*, 1900, pp. 253–97. A later version of it was published in *Archiv für Mathematik und Physik*, 3rd series, vol. 1 (1901), pp. 44–63 and 213–37, and this version was reprinted in David Hilbert, *Gesammelte Abhandlungen* III (Berlin 1935; reprinted New York 1965), pp. 290–329 [E.T. in *Bulletin of the American Mathematical Society* VIII (1902), pp. 437–79].

2 Evidently, the pagination of the offprint does not correspond to the pagination in the *Göttinger Nachrichten*. The passage Frege is referring to appears on p. 264: 'If our task is to investigate the foundations of a science, then we must first set up a system of axioms which contain a precise and complete description of those relations that obtain between the elementary concepts of that science.'

3 Cf. *Grundlagen der Geometrie*, sects 9ff.

reduced to the one I formulated and that it cannot therefore have a wider scope than mine. It would help to clear up matters if in your reply to my last letter – and I am still hoping for a reply – you could formulate such a principle precisely and perhaps elucidate its application by an example.

Since I am just now much concerned with the problem of irrationals, I am interested in the hint you give at the bottom of p. 13.[4] But I do not yet regard this kind of proof as workable, for two reasons one of which could perhaps be invalidated by the principle discussed above.

I have communicated our correspondnce to Dr Liebmann in Leipzig, who will lecture on the foundations of geometry in the winter semester. I hope you agree with this. Dr Liebmann will write me about his view as soon as he has formed an independent judgement.

With best regards,

Yours sincerely,
G. Frege

IV/8 [xv/8]   HILBERT to FREGE   22.9.1900

Göttingen
22 September 1900

Dear Colleague,

Many thanks for your letter, which I found upon my return from Aachen and which like your previous letters I have read with great interest. I know very well that I still owe you a detailed answer to your previous letter and that my debt to you is even greater now. But since I must again prepare a new lecture on partial differential equations in physics, I had better send you this card rather than nothing for the time being.

---

4 Frege is apparently referring to the following passage in Hilbert's lecture (ibid., p. 265):

Now I am convinced that we must succeed in finding a direct proof that arithmetical axioms are free from contradiction, if we carefully work through the known methods of inference in the theory of irrational numbers with that aim in view and try to modify them in a suitable manner.

In order to characterize the importance of the problem in yet another respect, I should like to add the following remarks. If a concept is assigned characteristic marks that contradict one another, then I say: the concept does not exist mathematically. Thus a real number whose square is equal to -1 does not exist mathematically. But if we succeed in proving that the characteristic marks assigned to the concept can never lead to a contradiction when subjected to a finite number of logical inferences, then I say that this proves the mathematical existence of the concept of, e.g., a number of function satisfying certain requirements. In the present case, where we are dealing with the axioms for real numbers in arithmetic, the demonstration that the axioms are free from contradiction is at the same time a proof of the mathematical existence of what is contained in the concept of real numbers or in that of the continuum.'

In my opinion, a concept can be fixed logically only by its relations to other concepts. These relations, formulated in certain statements, I call axioms, thus arriving at the view that axioms (perhaps together with propositions assigning names to concepts) are the definitions of the concepts. I did not think up this view because I had nothing better to do, but I found myself forced into it by the requirements of strictness in logical inference and in the logical construction of a theory. I have become convinced that the more subtle parts of mathematics and the natural sciences can be treated with certainty only in this way; otherwise one is only going around in a circle.

With best regards and in the hope that I may yet convince you and perhaps hear from you in spite of my (for the time being) hasty reply,

Yours sincerely,
Hilbert

IV/9 [xv/9]   HILBERT to FREGE   7.11.1903

Göttingen
7 November 1903

Dear Colleague,

Many thanks for the second volume of your *Basic Laws*, which I find very interesting.[1] Your example at the end of the book (p. 253)[2] was known to us here;*[3] I found other even more convincing contradictions as long as four or five years ago; they led me to the conviction that traditional logic is inadequate and that the theory of concept formation needs to be sharpened and refined.[4] As I see it, the most important gap in the traditional structure of logic is the assumption made by all logicians and mathematicians up to now that a concept is already there if one can state of any object whether or

---

* I believe Dr Zermelo discovered it three or four years ago after I had communicated my examples to him.

IV/9. 1 GGA II, published in 1903.
2 Hilbert is referring to Russell's antimony which is stated and discussed in detail in the postscript to GGA II.
3 E. Zermelo, at the time a lecturer in Göttingen, discovered Russell's antimony independently, according to accounts now available (cf. e.g. C. Reid, *Hilbert* (Berlin 1970), p. 98).
4 O. Blumenthal mentions in his biography of Hilbert that Hilbert became definitely convinced of the questionable nature of the mathematical discussions of infinity at the time 'by the example of the contradictory set of all sets that arises through combination and self-verification, an example he had thought up himself but never considered outside the context of purely mathematical operations'. (D. Hilbert, *Gesammelte Abhandlungen* III (Berlin 1935; reprinted New York 1965), pp. 421ff.)

not it falls under it. This does not seem adequate to me. What is decisive is the recognition that the axioms that define the concept are free from contradiction. I agree in general with your criticisms; except that you do not do full justice to Dedekind and especially Cantor.[5]

It is a pity that you were neither in Cassel nor in Göttingen;[6] perhaps you will decide to visit Göttingen between terms. Since rail travel is so comfortable today, personal communication is surely preferable to the written kind. I at least lack the time for the latter. There are a number of young scholars here interested in the 'axiomatization of logic'.

Yours sincerely,

Hilbert

-----

5 Cf. GGA II, sects 138ff for Frege's criticism of Dedekind and GGA II, sects 68ff for his criticism of Cantor.

6 The 75th Convention of German Scientists and Doctors was held in Cassel from 20 to 25 September 1903. The members of the German Mathematical Association who had gathered in Cassel subsequently visited the Institute for Mathematics and Physics in Göttingen (cf. *Jahresberichte der Deutschen Mathematiker-Vereinigung* 12 (1903), pp. 517 and 520). But Hilbert may also be referring to a session of the Göttingen Mathematical Society on 5 May 1903 at which E. Zermelo discussed Frege's work on the foundations of arithmetic (cf. ibid., pp. 345ff).

V FREGE–HÖNIGSWALD

Editor's Introduction

Richard Hönigswald (1875–1947), a neo-Kantian, was the editor of *Wissenschaftliche Grundfragen*, a series of philosophical monographs which appeared from 1924 on at irregular intervals. The whole correspondence concerns Frege's contribution to the series.

V/1 [xvii/4]   HÖNIGSWALD to FREGE   24.4.1925

Breslau
24 April 1925

My dear Colleague,

I still owe you many thanks for your penetrating paper, 'The Sources of Knowledge of Mathematics and the Mathematical Sciences', which reached me through your colleague Bauch in Jena. The spirit that informs your paper and the methodical results you reach correspond exactly to our endeavours and wishes. And it is only because our enterprise is *not* a philosophical *journal* which combines several papers in each number but a series of monographs, each 4 to 6 printer's sheets thick, which appear *independently* on the market, that I allow myself the humble request that you be so kind as to expand your paper on some especially important points and then resubmit it to us.

Without wanting to anticipate your decision in any way, I am taking the liberty, after a thorough study of your work, of indicating some points where an expansion seems especially worth considering.

p. 4 [269]:[1] You mention the paradoxes of set theory. Perhaps you would be so good as to go into them more deeply so as to provide further justification for your opinion that set theory is untenable. As things stand, this should have a special claim to being of current scientific interest.

p. 5 [270]: I should welcome it greatly if you could possibly give a more exact definition of the true concept of a function and the false one and characterize the superiority of the former in more detail. The same applies to

p. 8f [273f]: your subtle discussion of the old formulation of the axioms and the new one, and finally to

p. 9 [273]: the formal definition of the number concept and the informal one. These constitute an extremely interesting topic, on which you in particular, my dear Colleague, have on several occasions expressed yourself with such authority. Could you not let us have your views regarding the investigations carried out since then by Russell, Zermelo and Hilbert?

V/1. 1 The page numbers in brackets refer to PW.

So I very much hope, as I make this request, that we shall have the pleasure of receiving the expanded version of your paper as soon as your time and health permit.

Once more with many thanks and with all good wishes, I am,

Yours very sincerely,
R. Hönigswald

V/2 [xvii/5]   FREGE to HÖNIGSWALD   26.4–4.5.1925

Bad Kleinen, Mecklenburg
26 April 1925

My dear Colleague,

I really have more to thank you for than you me, for I am very anxious to have something from my pen appear in your *Wissenschaftliche Grundfragen*, especially since we agree so well on the main points. I find it of course perfectly in order that you should make some remarks and express some wishes in your letter of two days ago. I shall be sorry if I cannot quite satisfy your demands, which I find entirely justified, as a result of my state of health and the weakness of my memory.

28 April. I turn first to the paradoxes of set theory. They arise because a concept, e.g. fixed star, is connected with something that is called the set of fixed stars, which appears to be determined by the concept – and determined as an object. I thus think of the objects falling under the concept fixed star combined into a whole, which I construe as an object and designate by a proper name, 'the set of fixed stars'. This transformation of a concept into an object is inadmissible; for the set of fixed stars only seems to be an object; in truth there is no such object at all. For any object and any (first-level) concept there are only two possible cases: either the object falls under the concept or it does not fall under the concept. Applied to the present case this means: either the set of fixed stars falls under the concept fixed star or it does not fall under it, or in other words: either the set of fixed stars is a fixed star or it is not a fixed star.

2 May. Whatever is a fixed star is a member of the set of fixed stars. Thus if the set of fixed stars is a fixed star, then the set of fixed stars is a member of itself. Whatever is not a fixed star is not a member of the set of fixed stars. Thus if the set of fixed stars is not a fixed star, then the set of fixed stars is not a member of itself. We thus distinguish sets that are members of themselves from sets that are not members of themselves.

3 May. Let s be the set of sets that are not members of themselves. Is s a member of itself? Let us first suppose that it is. If something is a member of a set, then it falls under the concept that determines the set. Accordingly, if s is a member of itself, then s is a set that is not a member of itself. Our supposition that s is a member of itself thus leads to a contradiction. Let us suppose secondly that s is not a member of itself. Then s falls under the

concept that determines set *s* and is thus a member of it. Accordingly, if *s* is not a member of itself, then *s* is a member of itself. Thus the supposition that *s* is not a member of itself also leads to a contradiction. These are the paradoxes of set theory which make set theory impossible. Instead of 'set of *F*s', where '*F*' stands for a concept word, we could say just as well 'extension of *F*' or 'class of *F*s' or 'system of *F*s'. The essence of the procedure which leads us into a thicket of contradictions can be summed up as follows. The objects that fall under *F* are regarded as a whole, as an object, and designated by the name 'set of *F*s' ('extension of *F*', 'class of *F*s', 'system of *F*s', etc.). A concept word '*F*' is thereby transformed into the object name (proper name) 'set of *F*s'. This is inadmissible because of the essential difference between concept and object, which is indeed quite covered up in our word languages. Concept words and proper names are exactly fitted for one another.* Because of its need for completion (unsaturatedness, predicative nature), a concept word is unsaturated, i.e., it contains a gap which is intended to receive a proper name. Through such saturation or completion there arises a proposition whose subject is the proper name and whose predicate is the concept word, and which expresses that the object designated by the proper name falls under the concept.

4 May. In such a proposition, concept word and proper name occupy essentially different places, and it is obvious that a proper name will not fit into the place intended for the concept word. Confusion is bound to arise if a concept word, as a result of its transformation into a proper name, comes to be in a place for which it is unsuited. It is hardly possible to examine all expressions which language puts at our disposal for their admissibility. The expressions 'the extension of *F*' seems naturalized by reason of its manifold employment and certified by science, so that one does not think it necessary to examine it more closely; but experience has shown how easily this can get one into a morass. I am among those who have suffered this fate. When I tried to place number theory on scientific foundations, I found such an expression very convenient. While I sometimes had slight doubts during the execution of the work, I paid no attention to them. And so it happened that after the completion of the *Basic Laws of Arithmetic* the whole edifice collapsed around me. Such an event should be a warning not only to oneself but also to others. We must set up a warning sign visible from afar: let no one imagine that he can transform a concept into an object. So much for the paradoxes of set theory, at least for now. This exposition has become rather extensive. Originally, I pointed out the sources of knowledge without further justification. In its terse succinctness it was to have a hard-hitting effect; but you are quite right: the justification will still be missed; but it becomes

---

* We may here disregard expressions containing the words 'all' or 'some' and number words.

extensive, as these beginnings show. Would I not discourage some by continuing in this manner?

Unfortunately my state of health gives me much trouble. Please excuse the defects of this letter for this reason. If you think that this letter could somehow find a use in your *Wissenschaftliche Grundfragen* but would have to be reworked by me, please send it back to me.[1]

Yours very sincerely,
G. Frege

V/2. 1 It appears that Hönigswald returned this letter (or a copy of it) to Frege, for it was found in Frege's *Nachlass*, though it has since been lost and is now known only from a typewritten copy of Scholz's.

VI FREGE–HUNTINGTON

Editor's Introduction

Edward V. Huntington (1874–1952), the American mathematician, was professor of mechanics at Harvard from 1919 to 1941.

VI/1 [xviii/1]   FREGE to HUNTINGTON   undated[1]

Many thanks for kindly sending me 'A Complete Set of Postulates etc.', 'Complete Sets of Postulates etc.', and 'Simplified Definition of a Group'.[2] The first two were already known to me through the good offices of Professor Pyntzmer. I have not yet found the time to look deeply into the matter, but I have already seen enough to tell that your endeavours have some points of contact with my own and are therefore of great interest to me. I am just now busy with the printing of the second volume of my *Basic Laws of Arithmetic*, which contains some considerations similar to the ones in your papers, especially with regard to Archimedes' axiom and the commutative principle, even though our points of departure are different. I have set myself the goal of basing arithmetic on logic alone. For this it is essential to exclude with certainty everything derived from other sources of knowledge (intuition, sensible experience). And for this it is again necessary to produce for each proof a chain of inferences without gaps, so that every transition proceeds according to a known logical law. Thus nothing must be left to mere self-evidence, for its nature and laws are unknown. One could then never be certain that this evidence was purely logical. Thus it is not enough for me if a progression to a new thought is evident, but I also want to know by what laws it is justified. I first tried to carry out the proofs in the German language, but I soon became convinced that, carried out in this way, the proofs grew to an enormous length and that their strictness was nevertheless not ...[3] what I call endless, namely the number of all finite numbers, and show that endless is not a finite number. In the second volume, which is to appear shortly, I first carry the theory of numbers somewhat

---

VI/1. 1 The letter dates presumably from the year 1902, as indicated by the publication dates of Huntington's papers and Frege's GGA II.
2 The papers in question are: 'A Complete Set of Postulates for the Theory of Absolute Continuous Magnitude', *Transactions of the American Mathematical Society* 3 (1902), pp. 264–79; 'Complete Sets of Postulates for the Theories of Positive Integral and Positive Rational Numbers', ibid., pp. 280–4; 'Simplified Definition of a Group', *Bulletin of American Mathematical Society*, 2nd Series, 8 (1901–2), pp. 296–300.
3 The copy contains a gap at this point. For the missing material, cf. GGA I, sects 122ff.

further and also take into account infinite numbers. I then go on to real
numbers, and this is where the points of contact with your investigations are
to be found. I here raise the question, What properties must a class have in
order to be a domain of magnitudes? There are indeed some points where
my treatment also diverges from yours. I too take into account at once the
contrast between positive and negative, taking from Gauss the hint that this
contrast occurs only among relations. The question now arises in this form,
What properties must a class of relations have in order to be a domain of
magnitudes? This way of putting the question is somewhat simpler than
yours, because something corresponding to your 'rule of combination' is
given from the outset, namely the composition of relations, which was
already defined in the first volume. You take two things into account: the
class or 'assemblage' and the 'rule of combination' in this class, and this rule
is not given through the class. This is also a point where your account does
not seem to be perfectly correct logically.

As I said before, language is logically imprecise, and I have noticed this
especially with the definite article. Language tempts us to use it where it is
not really justified. The definite article serves to transform a concept word
(*nomen appellativum*) into a proper name (*nomen proprium*), I mean an
expression that is supposed to designate a certain object. Thus the definite
article is really justified only if it is established, first, that there is such an
object, and secondly, that there is no more than one. Now you write (p.
266): 'If the first of the two given elements is denoted by a and the second
by b, then the object which they determine is denoted by $a \circ b$.'[4] Here the
definite article in 'the object' seems to me incorrect. For what does '2 $\circ$ 3'
mean? We can assume very different rules of combination. If we take the
sum, then 5 is determined; if we take the product, 6 is determined; and so we
can assume countless further rules, each of which determines a different
number; but we cannot tell from '2 $\circ$ 3' which rule is to be assumed. I know
very well that you are not the first to use this sign. Others have made this
mistake before you; but I must nevertheless count it as a mistake. If the
signs contained some indication of the rule of combination, the matter might
be different; but as it is, '$a \circ b$' does not mean anything, and no combination
of signs of the form '$a \circ b$' means anything. It seems to me that we are here
presented with a cross between two different kinds of signs, namely those
that designate (or mean) and those that merely indicate, as I put it. I call the
signs '2', '3' proper names; each of them designates or means a definite
number. While the signs '=' and '+' are not proper names, each of them
nevertheless designates a certain relation. The sign '+' in '$a + b$' designates a
certain function with two arguments; it is, as I put it, a two-place function
sign. On the other hand, letters in general are indicative signs. 'a' does not

---

4 'A Complete Set of Postulates for the Theory of Absolute Continuous
Magnitude', loc. cit. ['$\circ$' is intended as a variable.]

designate a definite object but only indicates, and similarly for the letter 'f', as in '$f(2, 3)$'. Such sign combinations as '$a + b$', '$f(a, b)$' do not therefore mean anything and have no sense by themselves, but can help to express a sense in the context of a proposition, as in '$a + b = b + a$' or 'if $a + b = c$, then $a = c - b$' and similar ones. By making use here of indicative letters instead of designating words, we impart generality to thoughts. Now the sign 'O' seems to me to be a hybrid, for it is not a letter, and the way it is introduced makes us take it for a designating one. But we later see that it does not designate anything, but is used more like an indicative sign. Its hybrid nature gives rise to unclarities about the sense of the propositions in which it occurs. Is it to impart generality to thoughts like an indicative sign? And what are the component parts of such a thought? If O is supposed to be an indicative sign, well, let us replace it by a letter and write '$f(a, b)$' instead of 'a O b'. This obliges us at once to give more precise limits to the thought to which we wish to impart generality by using the sign 'f', and the whole account gains in definiteness. We should not be deterred by the fact that as the corresponding arguments we must take not just numbers but objects as such. I used function letters in this way in my *Conceptual Notation* as early as 1879.

VII FREGE–HUSSERL

Editor's Introduction

Edmund Husserl (1859–1938) corresponded with Frege at two different times: in 1891 while he was in Halle and in 1906 while he was at Göttingen. In both cases, the occasion was Husserl's sending Frege copies of some of his writings. In 1891 the writings included Husserl's *Philosophie der Arithmetik*,[1] which Frege was to review in 1894, and in return, Husserl received Frege's *Begriffsschrift*[2] as well as the other writings mentioned in their letters. Husserl knew some of Frege's other writings as well, as indicated by underlining and marginal comments in the copies or offprints found in his *Nachlass*. Besides, his *Nachlass* contained a complete collection of Frege's writings between 1891 and 1894.[3] It is especially noteworthy that this collection was broken off after Frege's review of Husserl's *Philosophie der Arithmetik* and resumed only several years later with Frege's series of articles on the foundations of geometry published between 1903 and 1906. This confirms the hypothesis that Frege's review was responsible for a 'cooling off' in what had evidently been a good relationship.[4] But even during this period, Husserl continued to study Frege's views. There exist excerpts made by him from the correspondence between Frege and Hilbert in 1899 and 1900. They are appended to the manuscript of his lecture on 'The Imaginary in Mathematics', which he delivered in 1901 to the Göttingen Mathematical Society. Husserl probably owed these excerpts to Hilbert, whose colleague he became in 1901.[5]

In view of the fact that Husserl and Frege had fairly precise knowledge of each other's writings, it is surprising that they hardly refer to each other, except for Husserl's *Philosophie der Arithmetik* and Frege's review of it. Some recent authors have tried to clear up the relationship between them.[6] There is now general

---

1 *Philosophie der Arithmetik: Psychologische und logische Untersuchungen* I (Halle 1891).

2 Husserl's notes on *Begriffsschrift* are published in *Begriffsschrift und andere Aufsätze* ⟨30⟩, pp. 117–21 (Appendix II).

3 Cf. I. Angelelli's comments in *Kleine Schriften* ⟨31⟩, pp. 431–4 (Appendix III). Husserl's marginal notes to 'Function und Begriff' are published ibid., pp. 433–4.

4 Cf., e.g., Henry Pietersma, 'Husserl and Frege', *Archiv für Geschichte der Philosophie* II (1967), pp. 298–323, esp. 298–9.

5 These excerpts are published in *Philosophie der Arithmetik, mit ergänzenden Texten (1890–1901)*, Lothar Eley, ed. (The Hague 1970) (Husserliana vol. XII), pp. 447–51.

6 Cf. the following works:

(1) Lothar Eley, *Metakritik der Formalen Logik* (The Hague 1969) (Phaenomenologica 31);

(2) Dagfinn Føllesdal, *Husserl und Frege: Ein Beitrag zur Beleuchtung der Entstehung der phänomenologischen Philosophie* (Oslo 1958);

(3) Günter Mortan, 'Einige Bemerkungen zur Überwindung des Psychologismus durch G. Frege und E. Husserl,' *Atti del Congresso Internazionale di Filosofia 1958* (Florence 1961), XII, pp. 327–34;

(4) Andrew D. Osborn, *Edmund Husserl and his 'Logical Investigations'*, 2nd ed. (Cambridge, Mass., 1949);

(5) Henry Pietersma, 'Husserl and Frege', *Archiv für Geschichte der Philosophie* II (1967), pp. 298–323;

(6) Robert C. Solomon, 'Sense and Essence: Frege and Husserl', *International Philosophical Quarterly* X (1970), pp. 378–401.

agreement that Frege's critique of the basic psychologistic attitude in *Philosophie der Arithmetik* prepared the way for Husserl's rejection of psychologism and may even have occasioned it. Some even regard Frege as the true author of the refutation of psychologism and refer in this connection to Frege's arguments in the preface to *Grundgesetze der Arithmetik* I.[7] Husserl himself only makes the following admission in a footnote to his *Logische Untersuchungen*: 'I need hardly say that I no longer approve of the criticisms *of principle* that I levelled against Frege's anti-psychologistic position in my *Philosophie der Arithmetik* I, pp. 129–32.'[8] And here Husserl too refers to the preface to *Grundgesetze der Arithmetik* I. But he does not seem conscious of having been influenced by Frege, or perhaps he would not admit it. At any rate, later, in 1936, he writes the following memorable lines to Scholz: 'I never got to know G. Frege personally, and I no longer remember the occasion for our correspondence. At the time he was generally regarded as an outsider who had a sharp mind but produced little or nothing, whether in mathematics or in philosophy.'[9]

VII/1 [xix/1]   FREGE to HUSSERL   24.5.1891

Jena
24 May 1891

My dear Doctor,

By sending me your *Philosophy of Arithmetic*,[1] as well as your critical notice of Schröder's *Lectures on the Algebra of Logic*[2] and your essay on 'The Inferential Calculus and The Logic of Content',[3] you have given me much pleasure, the more so as I myself have been much concerned with such questions. Besides divergences from my views, I believe I also see some points of agreement in your writings. I have read your notice of Schröder's work with great interest, and it has prompted me to write down my own thoughts now, rather than wait for the second volume to appear, as I had planned to do earlier. Perhaps my notes will appear in *Zeitschrift für*

---

7 Cf. G. Mortan, op. cit., p. 329.
8 *Logische Untersuchungen* I and II (Halle 1900–1), vol. I, p. 169 [E.T. by J. N. Findlay, *Logical Investigations* (London and New York 1970), vol. I, p. 179].
9 Postcard dated 19.2.1936.
VII/1. 1. *Philosophie der Arithmetik; Psychologische und logische Untersuchungen* I (Halle 1891).
2 Review of Ernst Schröder, *Vorlesungen über die Algebra der Logik* (*Exakte Logik*) I (Leipzig 1890), in *Göttingsche gelehrte Anzeigen* (1891), pp. 243–78. Ernst Schröder (1841–1902) was professor of mathematics, first at the College of Science and Technology in Darmstadt and, from 1876 on, at the College of Science and Technology in Karlsruhe. His works include *Der Operationskreis des Logikkalkuls* (Leipzig 1877), and *Vorlesungen über die Algebra der Logik* (*Exakte Logik*) I–III (Leipzig 1890—1905).
3 'Der Folgerungscalcul und die Inhaltslogik', *Vierteljahrsschrift für wissenschaftliche Philosophie* XV (1891), pp. 168–89, with Supplement, pp. 351–6.

Philosophie und philosophische Kritik.[4] I agree with you when you find that Schröder's definitions of 0, 1, $a + b$, and $a \cdot b$ are faulty.[5] Strictly speaking, instead of defining $a + b$, Schröder defines the sign $\in$ a second time together with the sign $+$. One and the same definition must not be used to define two different things, and least of all, something that had already been defined previously. The sign $\in$ should have been introduced at once for all possible cases. Even if $a + b$ was then defined by itself, it should follow without further ado what was to be understood by '$a + b \in c$'. If we clutch our foreheads in astonishment at the first appearance of $\sqrt{-1}$ and ask what this means, that is only a sign that the first introduction of the root sign contained a gap and was therefore faulty. It should have been such that there could be no doubt about $\sqrt{-1}$, after the minus sign and '1' had also been explained correctly. As far as Schröder's argument on p. 245 is concerned, which you call a sophism,[6] I believe I can show that from Schröder's principles we can indeed get both Schröder's result and yours, because his 'class' combines two very different meanings in one.[7] If we hold on to one of them, then we must absolutely reject Schröder's 0; the algorithm can incidentally be maintained if the meaning of '$\in$' is defined

---

4 Frege completed this project, but his paper $\langle 13 \rangle$ appeared instead in *Archiv für systematische Philosophie*.

5 Husserl's criticism (loc. cit., pp. 267–9) of Schröder's definitions of '0', '1', '$a + b$', and '$a \cdot b$' is to the effect that Schröder neglected to demonstrate that these definitions are justified. To do this, one needs to do more than demonstrate that they are free from contradiction. One must also show that after the addition of these signs the calculus remains a calculus of sets, i.e., that it does not come into 'conflict' with the original intentions. Here Husserl objects in general to the 'formalistic' conception of symbols, according to which symbols are given an 'interpretation' only 'after the fact', and also against so-called 'creative definitions'. In this Husserl shares Frege's view, which Frege had already expressed in *Ueber formale Theorien der Arithmetik* $\langle 6 \rangle$. Frege himself criticizes Schröder's definitions mainly on the ground that they violate the principles of 'completeness' and 'simplicity', which Frege was to name and discuss later in GGA II (sects 55–67). The definition in question is to be found in Schröder's *Vorlesungen über die Algebra der Logik* I, p. 196. According to it, '$a + b \in c$' is defined as '$a \in c$' and '$b \in c$'. The sign '$\in$', which is composed of '$\subset$' and '$=$', stands for subordination, including the improper kind. Schröder introduces '$a \cdot b$' (or for short 'ab') by defining '$c \in ab$' as '$c \in a$' and '$c \in b$'. To put it informally, the logical sum '$a + b$' stands for the *combination* of 'a' and 'b', and the logical product '$a \cdot b$' stands for the *overlapping portion* of 'a' and 'b'. The signs '0' and '1' are defined respectively through the 'general validity' of '$0 \in a$' and '$a \in 1$' and thus stand for the null-class and the universal class respectively.

6 This objection of Husserl's (loc. cit., p. 272) refers to Schröder's demonstration of the contradictoriness of Boole's concept of 1 as the class of all conceivable things (universe of discourse). To avoid these contradictions, Schröder develops in his lectures a kind of theory of types (I, pp. 245–9). Cf. A. Church, 'Schröder's Anticipation of the Simple Theory of Types', *Journal of Unified Science* IX (1939), pp. 149–52 and W. V. O. Quine's review of this in *Journal of Symbolic Logic* V (1940), p. 71.

7 This is explained in detail in $\langle 13 \rangle$, pp. 439ff.

accordingly, but this has nothing whatsoever to do with logic. In the other case 0 is admissible; and we are really engaged in logical considerations; but intuitive clarity is missing, and Euler's diagrams are not altogether unexceptionable.[8]

But I can only hint at this here.

I thank you especially for your *Philosophy of Arithmetic*, in which you take notice of my own similar endeavours, perhaps more thoroughly than has been done up to now. I hope to find some time soon to reply to your objections. All I should like to say about it now is that there seems to be a difference of opinion between us on how a concept word (common name) is related to objects. The following schema should make my view clear:

Proposition	proper name	concept word		
↓	↓	↓		
sense of the proposition (thought)	sense of the proper name	sense of the concept word		
↓	↓	↓		
meaning of the proposition (truth value)	meaning of the proper name (object)	⎛ meaning of the concept word (concept) ⎞	→	object falling under the concept

With a concept word it takes one more step to reach the object than with a proper name, and the last step may be missing – i.e., the concept may be empty – without the concept word's ceasing to be scientifically useful. I have drawn the last step from concept to object horizontally in order to indicate that it takes place on the same level, that objects and concepts have the same objectivity (see my *Foundations*, sect. 47). In literary use it is sufficient if everything has a sense; in scientific use there must also be meanings. In the *Foundations* I did not yet draw the distinction between sense and meaning. In sect. 97 I should now prefer to speak of 'having a meaning' instead of 'having a sense'. Elsewhere, too, e.g. in sects 100, 101, 102, I would now often replace 'sense' by 'meaning'. What I used to call judgeable content is now divided into thought and truth value.[9] Judgement in the narrower sense could be characterized as a transition from a thought to a truth value.

Now it seems to me that for you the schema would look like this:

concept word

↓

sense of the concept word
(sense)

↓

object falling under the concept

8 Euler's diagrams, so called after the mathematician Leonhard Euler (1707–83), represent logical relations between extensions (classes) by means of spatial relations between circles.

9 Cf., e.g., GLA, p. 77, note **. In 'Begriff und Gegenstand', ⟨10⟩, p. 198, Frege refers specifically to this place. Cf. also BS, p. 2.

so that for you it would take the same number of steps to get from proper names to objects as from concept words. The only difference between proper names and concept words would then be that the former could refer to only one object and the latter to more than one. A concept word whose concept was empty would then have to be excluded from science just like a proper name without a corresponding object.

In return for your valuable gifts I can unfortunately send you for now only a few shorter writings. From the two lectures on my conceptual notation[10] you should be able to tell that Schröder's judgement (in his *Algebra of Logic*, p. 95 n.) is unfounded.[11] It is true that these lectures do not accurately reflect my present position, as you can see by comparing them with 'On Function and Concept'. But since they can easily be translated into my present terminology, they may nevertheless serve to give you an idea of the use of my conceptual notation. Instead of speaking of a 'circumstance', one should speak of a 'truth value'.

In the hope that the exchange of ideas between us will be continued and that it will contribute something to the advancement of science, I remain,

Yours sincerely,

Dr G. Frege

VII/2 [xix/2]   HUSSERL TO FREGE   18.7.1891

Halle

18 July 1891

My dear Professor,

Forgive me for not thanking you till today for your friendly letter as well as the papers you sent me. I did not want to thank you before I had studied your papers and was able to continue the substantial discussion you initiated in your letter. Unfortunately, the circumstances were not favourable to my wishes: I hoped in vain to find enough free time for what seemed to me necessary first of all: to form a clear picture of the nature and scope of your original conceptual notation. In this respect I cannot expect very much from the next few weeks either, and so I will not delay my answer any longer.

First of all, allow me to acknowledge the large amount of stimulation and encouragement I derived from your *Foundations*. Of all the many writings that I had before me when I worked on my book, I could not name another which I studied with nearly as much enjoyment as yours. Although I could not on the whole agree with your theories, I derived constant pleasure from

---

10 ⟨2⟩ and ⟨3⟩.

11 In *Vorlesungen über die Algebra der Logik* I, we read (ibid.): 'I believe I have shown in my review that Mr Frege's "concept notation" does not deserve this name but would have to be called a logical "judgement notation" (which, though logical, is not the most appropriate one).'

the originality of mind, clarity, and, I should almost like to say, honesty of your investigations, which nowhere stretch a point or hold back a doubt, to which all vagueness in thought and word is alien, and which everywhere try to penetrate to the ultimate foundations.

As will be evident from this, I have long cherished the wish to get to know the rest of your works, and I am very grateful to you for sending me a series of them and in particular for making it possible for me to study one of them which I tried in vain to get hold of ('On Formal Theories of Arithmetic').[1]

I too have noticed that in spite of fundamental divergences our views touch on several points. Some observations that forced themselves upon me I find expressed by you many years earlier. For example, compare your excellent remark ('On the Purpose of the Conceptual Notation', p. 1):[2] 'For the problems dealt with by Boole seem in large part to have been invented in order to be solved by means of his formulas' with my review, p. 278.[3] Also the fundamental distinction between a *language* and a *calculus*, which I stress on p. 258, is already drawn by you, ibid., p. 2, where you separate the concepts *calculus ratiocinator* and *lingua characterica*. It does seem to me, though, that since the conceptual notation is supposed to be a *lingua characterica*, it should not be called 'a formal language modelled on the arithmetical one'.[4] For one thing is surely certain, and this is that arithmetic is a *calculus ratiocinator* and not a *lingua characterica*.

I have not yet read Schröder's criticism of your conceptual notation. I can well understand it if, as you say, he does not do justice to you; he lacks what is indispensable for investigations in the area we are talking about: a fine and acute logical mind. His strength lies in a very different direction: he is a brilliant arithmetical technician, but no more than that.

Unfortunately I have not yet been able to make full use of the contents of your communications, perhaps because I still lack a thorough knowledge of your pertinent writings. Thus I still have only a rough idea of how you want to justify the imaginary in arithmetic. After many vain efforts I found a path leading to that goal, but in your discussion of it ('On Formal Theories of Arithmetic', p. 8),[5] you hold that path to be impassable. During the summer vacations I am thinking of turning my relevant drafts into a clear copy. Perhaps I shall then succeed in summarizing the essentials of my theory in a brief letter.

VII/2. 1 ⟨6⟩.

2 ⟨3⟩.

3 Husserl there writes: 'As things now stand, the representatives of the new discipline are struggling with the peculiar difficulty of trying to discover problems to be solved by their beautiful methods, and they nowhere show a greater sharpness of mind than in the construction of illustrations and examples, which look artificial enough as a result.'

4 This is the subtitle of BS.

5 The page reference is to the offprint. In the published version, p. 8 corresponds to p. 101, l. 10–p. 102, l. 9 [E.T. pp. 149–50].

I am completely at one with you in your rejection of 'formal arithmetic', as it is now usually presented: not merely as an extension (and certainly a very significant one) of arithmetical techniques, but as a *theory* of arithmetic. The much-praised book by Hankel, which is quite confused from a logical point of view, has created much confusion in this respect.[6] From everything I hear, the English writers seem to be much clearer, especially Peacock.[7] Unfortunately I was unable to obtain Peacock's famous *Algebra* (1845). The same for Gregory's writings.[8] I recently found some hints about their theories in a quite valuable little book by Fine (*The Number System of Algebra*, Boston and New York 1891).[9]

Yours very sincerely,
Dr E. G. Husserl

VII/3 [xix/3]   FREGE to HUSSERL   30.10–1.11.1906

Jena
30 October to 1 November 1906

Dear Colleague,

As I thank you very much for kindly sending me your article,[1] may I at the same time pass on to you some observations that occurred to me as I was reading it, since I do not have the time now to go into it thoroughly.

---

6 Hermann Hankel (1839–73), *Vorlesungen über die complexen Zahlen und ihre Functionen*, in two parts; part I: *Theorie der complexen Zahlensysteme, insbesondere der gemeinen imaginären Zahlen und der Hamilton'schen Quaternionen nebst ihrer geometrischen Darstellung* (Leipzig 1867). Hankel became especially well known through his proof of the proposition that an expansion of the complex number system is impossible if the laws of arithmetic are to remain valid.

7 George Peacock (1791–1858). Husserl is referring to his work, *A Treatise on Algebra* I–II (London 1842–5). Peacock was the first to formulate the so-called 'principle of permanence' (principle of the permanence of equivalent forms) in 'Report on the Recent Progress and Present State of Certain Branches of Analysis', *Report of the Third Meeting of the British Association for the Advancement of Science, held in 1833*, pp. 185–353, esp. pp. 198–207.

8 Duncan Farquharson Gregory (1813–44), Fellow of Trinity College, Cambridge, and founder (in 1839) of the *Cambridge Mathematical Journal*, where many of his articles appeared. His other works are: *Essay on the Foundation of Algebra* (Edinburgh 1838); 'On the Real Nature of Symbolical Algebra', *Transactions of the Royal Society of Edinburgh* XIV (1840), pp. 208–16; and *Collection of Examples of the Processes of the Differential and Integral Calculus* (Cambridge 1841).

9 Henry Burchard Fine (1858–1928), from 1890 on professor of mathematics at Princeton University. Fine studied in Leipzig (1884–5) and obtained his doctorate there (1885).

VII/3. 1 The article in question is the fifth and concluding article in the series *Bericht über deutsche Schriften zur Logik in den Jahren 1895–99*, in *Archiv für systematische Philosophie* X (1904), pp. 101–25. There is no way of telling whether Frege also received the other articles. The fifth article is a critical discussion of Anton Marty, *Über subjektlose Sätze und das Verhältnis der Grammatik zur Logik und Psychologie*, 6th and 7th articles, *Vierteljahrsschrift für wissenschaftliche Philosophie* XIX (1895), pp. 19–87 and 263–334.

Logicians make many distinctions between judgements which seem to me immaterial, and on the other hand they do not make many distinctions which I regard as important. It seems to me that logicians still cling too much to language and grammar and are too much entangled in psychology. This is apparently what prevents them from studying my conceptual notation, which could have a liberating effect on them. They find that my conceptual notation does not correctly represent mental processes; and they are right, for this is not its purpose at all. If it occasions entirely new mental processes, this does not frustrate its purpose. Apparently it is still thought to be the task of logic to study certain mental processes. Logic has really no more to do with them than with the movements of celestial bodies. It is in no sense part of psychology. Pythagoras' theorem expresses the same thought for all men, whereas everyone has his own images, feelings and decisions, different from everyone else's. Thoughts are not mental entities, and thinking is not an inner generation of such entities but the grasping of thoughts which are already present objectively. One should make only those distinctions with which the laws of logic are concerned. In gravitational mechanics no one would want to distinguish bodies according to their optical properties. Object and concept are not distinguished at all or much too little. Of course, if they are both ideas in the psychological sense, the difference is hardly noticeable. This is connected with the distinction between first- and second-level concepts, which is very important, but who among logicians knows anything about it? In logic, one must decide to regard equipollent propositions as differing only according to form. After the assertoric force with which they may have been uttered is subtracted, equipollent propositions have something in common in their content, and this is what I call the thought they express. This alone is of concern to logic. The rest I call the colouring and the illumination of the thought. Once we decide to take this step, we do away at a single stroke with a confused mass of useless distinctions and with the occasion for countless disputes which cannot for the most part be decided objectively. And we are given free rein to pursue proper logical analyses. Judged psychologically, the analysing proposition is of course always different from the analysed one, and all logical analysis can be brought to a halt by the objection that the two propositions are merely equipollent, if this objection is indeed accepted. For it will not be possible to draw a clearly recognizable limit between merely equipollent and congruent propositions. Even propositions which appear congruent when presented in print can be pronounced with a different intonation and are not therefore equivalent in every respect. Only now that logical analysis proper has become possible can the logical elements be recognized, and we can see the clearing in the forest. All that would be needed would be a single standard proposition for each system of equipollent propositions, and any thought could be communicated by such a standard proposition. For given a standard proposition everyone would have the whole system of equipollent propositions, and he could make the transition to any one of them whose illumination was particularly to his taste. It cannot be the task of logic to investigate language and determine what is contained in a linguistic expression. Someone who wants to learn logic from language is like an adult who wants to learn how to think from a child. When men created language, they were at the stage of childish pictorial thinking. Languages are not made so as to match logic's ruler. Even the logical element in language seems hidden behind pictures that are not always accurate. At an early time in the creation of language there occurred, it seems, a tremendous exuberance in the growth of linguistic forms. At a later time much of this had to be got rid of again and simplified. The main task of the logician is to free himself from language and to simplify it. Logic should be the judge of languages. We should either tidy up logic by throwing out subject and predicate or else restrict these concepts to the relation of an object's falling under a concept (subsumption). The relation of subordination of one concept under another is so different from it that it is not permissible to speak of subject and predicate also in this case.

With regard to propositions combined by 'and' and 'neither ... nor' (p. 121)[2] I am essentially in agreement with you. I would put it like this: The combination of two propositions by 'and' corresponds to the combination of two thoughts into one thought, which can be negated as a whole and also recognized to be true as a whole.

With regard to the question whether the proposition 'If A then B' is equipollent to the proposition 'It is not the case that A without B', one must say the following.[3] In a hypothetical construction we have as a rule improper propositions of such a kind that neither the antecedent by itself nor the consequent by itself expresses a thought, but only the whole propositional complex. Each proposition is then only an indicative component part, and

---

2 The page reference is to Husserl's fifth article cited in note 1 above. Husserl there criticizes Marty's view that in the case of statements combined by 'and' and 'neither ... nor' the affirmation or negation does not extend uniformly to the whole statement but is divided up between each of the two partial statements (cf. Marty, op. cit., p. 300).

3 Husserl (ibid., pp. 121ff) criticizes Marty's view (ibid., p. 304, note) that the two propositions are 'identical in sense'. Besides denying that they have 'congruence', Husserl also denies them 'equivalence (equipollence)'. The latter is present for Husserl if the negation of the two propositions also yields again 'something equivalent'. Now according to Husserl, the negation of 'If A then B' yields 'A can hold without B holding', and the negation of 'It is not the case that A (A does not hold) without B' (as Husserl puts it with reference to Marty) yields 'A *holds* without B holding'. This is why the two propositions are not equipollent for Husserl. In present-day terms, Husserl's analysis differs from Frege's subsequent analysis in that Frege regards 'if ... then' as defined as *material* implication, whereas Husserl appears to mean *strict* implication. In terms of strict implication, 'if ... then' can be explained as 'It is not possible that A holds without B holding'. The negation of this yields 'It is possible that A holds without B holding'. If this is replaced by 'It may be that A holds without B holding', then we can also use instead Husserl's formulation above: 'A may hold without B holding'.

each proposition indicates the other (*tot ... quot ...*). In mathematics, such component parts are often letters (If $a > 1$, then $a^2 > 1$). The whole proposition thereby acquires the character of a law, namely generality of content. But let us first suppose that the letters 'A' and 'B' stand for proper propositions. Then there are not just cases in which A is true and cases in which A is false; but either A is true or A is false; *tertium non datur*. The same holds for B. We then have four combinations:

A is true and B is true,

A is true and B is false,

A is false and B is true,

A is false and B is false.

Of these the first, third, and fourth are compatible with the proposition 'If A then B', but not the second. We therefore obtain by negation:[4] A is true and B is false, or: A holds without B holding, just as on the right-hand side.[5]

Let us suppose in the second place that the letters 'A' and 'B' stand for improper propositions; then it is better if we replace 'A' and 'B' by '$\Phi(a)$' and '$\Psi(b)$', where 'a' is the indicative component part. The proposition 'If $\Phi(a)$ then $\Psi(a)$' now has generality of content, and its negation cancels this generality and says that there is an object (say Δ) such that $\Phi(\Delta)$ is true and $\Psi(\Delta)$ is false. This is presumably what you mean by the words 'A may hold without B holding'.[6] The proposition '$\Phi(a)$ does not hold without $\Psi(a)$ holding' is now understood as follows: 'In general, whatever a may be, $\Phi(a)$ does not hold without $\Psi(a)$'. By negation we obtain: 'It is not in general so that, whatever a may be, $\Phi(a)$ does not hold without $\Psi(a)$'. In other words: 'There is at least one object (say Δ) such that $\Phi(\Delta)$ is true while $\Psi(\Delta)$ is false.' We get the same as on the left-hand side.[7] In each case we therefore

---

4 'Negation' is here applied to 'If A then B'. Thus Frege applies here the procedure, mentioned by Husserl, of comparing the negations of the two propositions.

5 We must imagine that two propositions 'If A then B' and 'It is not the case that A without B' as combined into an equation, whose left side is (at first) the proposition 'If A and B' and whose right side is (at first) the proposition 'It is not the case that A without B'. This corresponds to the equation used by Husserl (ibid., p. 121). If, like Frege, we interpret the left-hand side as a material implication, then the two sides agree after negation according to Frege's analysis.

6 This assumption of Frege's is only partly correct, as shown by Husserl's note 11, ibid., p. 122. For Husserl there distinguishes the cases in which A and B 'mean' propositions and those in which they 'mean' concepts, and hence just those cases that Frege wants to distinguish here. Accordingly, Husserl does not, contrary to Frege's assumption, reserve the form with 'may' for the latter case but also extends it to the former case, which must then be interpreted as strict implication. The latter case is also called 'formal implications' after Whitehead and Russell (*Principia Mathematica* I, pp. 20ff).

7 Cf. note 5, above.

have an equipollence. If we consult my *Conceptual Notation*, which is now already 28 years old, we find the answer to such a question without further ado. Now are these propositions also congruent? This could well be debated for a hundred years or more. At least I do not see what criterion would allow us to decide this question objectively.

But I do find that if there is no objective criterion for answering a question, then the question has no place at all in science.

Yours sincerely,

G. Frege

VII/4 [xix/6]   FREGE to HUSSERL   9.12.1906

Jena

9 December 1906

Dear Colleague,

Thank you very much for your letter of 16 November,[1] which prompts me to make the following remarks.

It seems to me that an objective criterion is necessary for recognizing a thought again as the same, for without it logical analysis is impossible. Now it seems to me that the only possible means of deciding whether proposition *A* expresses the same thought as proposition *B* is the following, and here I assume that neither of the two propositions contains a logically self-evident component part in its sense. If *both* the assumption that the content of *A* is false and that of *B* true *and* the assumption that the content of *A* is true and that of *B* false lead to a logical contradiction, and if this can be established without knowing whether the content of *A* or *B* is true or false, and without requiring other than purely logical laws for this purpose, then nothing can belong to the content of *A* as far as it is capable of being judged true or false, which does not also belong to the content of *B*; for there would be no reason at all for any such surplus in the content of *B*, and according to the presupposition above, such a surplus would not be logically self-evident either. In the same way, given our supposition, nothing can belong to the content of *B*, as far as it is capable of being judged true or false, except what also belongs to the content of *A*. Thus what is capable of being judged true or false in the contents of *A* and *B* is identical, and this alone is of concern to

VII/4. 1 Husserl's letter of 16 November is lost, together with another letter of 10 November. These two letters contained Husserl's reply to Frege's letter of 30 October to 1 November. According to Scholz, the first letter dealt with 'equipollent propositions and "colouring"' as well as logic in general, while the second letter dealt with 'the paradoxes', possibly Russell's paradox. On 21 December 1906 to 13 January 1907 Husserl wrote another two-part letter to Frege which is also lost. The first part was a continuation of Husserl's letter of 16 November and contained remarks on 'the paradoxes' and 'hypothetical structures'. The second part was a reply to the present letter.

logic, and this is what I call the thought expressed by both *A* and *B*. One can indeed count many sorts of things as part of the content of *A*, e.g., a mood, feelings, ideas; but none of these is judged true or false; at bottom it is of no concern to logic, just as whatever is incapable of being judged morally good or bad is of no concern to ethics. Is there another means of judging what part of the content of a proposition is subject to logic, or when two propositions express the same thought? I do not think so. If we have no such means, we can argue endlessly about logical questions without result.

I have further doubts about the following. You write, 'The form containing "all" is normally so understood that the existence of objects falling under the subject and predicate concepts is part of what is meant and is presupposed as having been admitted.' It seems to me that you can only give this the sense you want it to have if you strike out the words 'part of what is meant'. For if existence was part of what was meant, then the negation of the proposition 'all *m* are *n*' would be 'there is an *m* that is not *n*, or there is no *m*'. But it seems to me that this is not what you want. You want existence to be presupposed as having been admitted, but not to be part of what is meant. Now I use the expressions containing 'all' in such a way that existence is neither part of what I mean nor something I presuppose as having been admitted. Linguistic usage cannot be absolutely decisive here, since we need not be concerned with what linguistic usage is. Instead, we can lay down our linguistic usage in logic according to our logical needs. The reason for the usage I have laid down is simplicity. If a form of expression, like the one containing 'all', is to be used as a fundamental form in logical considerations, it is not feasible to use it so as to express two distinguishable thoughts at the same time, unless the proposition consists of two propositions combined by 'and'. For one must always strive to go back to the elements, to the simple. It must be possible to express the main thought without incidental thoughts. This is why I do not want the incidental thought of existence to be part of what I mean when I use an expression containing 'all'.

As always,

Yours sincerely,
G. Frege

Editor's Introduction

Philip E. B. Jourdain (1879–1919) published extensively on mathematics and the natural sciences, and was regarded for a long time as the leading authority on Newton. At the same time he carried on an extensive correspondence with important mathematicians throughout the world. He was co-editor (for England) of the *International Journal of Ethics* and, from 1912 on, of the *Monist*. In 1918, after the death of Paul Carus, he became editor-in-chief of the latter journal.

On Jourdain's life, see the obituary signed D.M.W. in *Proceedings of the London Mathematical Society*, 2nd Series 19 (1920), pp. lix–lx, and further: George Sarton, 'Philip E. B. Jourdain (1879–1919)', *Isis* 5 (1923), pp. 126–9; Laura Jourdain, 'Some Reminiscences on the Moral Personality of Philip E. B. Jourdain', ibid., pp. 129–33; as well as the obituary 'Philip Edward Bertrand Jourdain' in the *Monist* 30 (1920), pp. 161–82, part of which (pp. 162–71) consists of reminiscences by Jourdain's sister Millicent.

Jourdain's account of Frege's logical and mathematical theories (partly based on this correspondence) is printed as an appendix to this volume.

Jourdain's letters are here given in the original English (hence some divergence from the normal style of this volume). Frege invariably replied in German.

VIII/1 [xxi/1]   JOURDAIN to FREGE   7.9.1902

> 63, Chesterton Rd.
> Cambridge
> England
> Sept. 7. 1902

Dear Sir

Through a reference in Dedekind: 'Was sind u. was sollen die Zahlen?' 1893, S. XVII, I was led to your 'Grundlagen der Arithmetik', 1884. As I am engaged in writing a historical account of the theory of Manifolds (G. Cantor, Dedekind, ...), and your most stimulating book seems to contain views of the number-concept which, to a great extent, anticipate the works of others* which are connected with the theory named, I venture to ask you for some details which will help me to give a better account of your book.

From the example on p. 103, I gather that fundamental laws of arithmetic (e.g.,; if m and y are numbers, either $y = ,>$, or $< m$), stated in a form which *presupposes* the concept of the series of integers given in the 'Grundlagen', were already proved in your 'Begriffsschrift' 1879. I thus suspect that the 'Begriffsschrift' *led* you to the theory in your later (1884) book, (and, as such, the 'Begriffsschrift' should be analysed to the best of my ability in my

---

*I think in especial of Dedekind and the Italian 'logical school' (Peano, Burali-Forti, ...)

historical account). Will you kindly tell me whether this surmise is correct or not? Also, I should like to have the titles of any other of your works which bear on the subject in question.

Yours sincerely
Philip E. B. Jourdain

VIII/2 [xxi/2] FREGE to JOURDAIN 23.9.1902

Jena
23 September 1902

Dear Mr Jourdain,

Upon my return from a journey I found your letter of 7 September.

The proposition 'If m and y are whole numbers, then y is equal to or greater or smaller than m' is a special case of the proposition I proved in my *Conceptual Notation* about objects as such arranged in serial order by some uniquely ordering relation. I expressed and proved this law in a somewhat different way in formula 243 of my *Basic Laws of Arithmetic* I (Jena, Pohle, 1893). I believe I have perfected my conceptual notation somewhat in this work (of which vol. II is to appear shortly), and I regard it in many respects as better than Peano's, even though it may appear less simple at first glance. In a letter to me Mr Bertrand Russell called my attention to the fact that my fundamental law V is in need of restriction. But our correspondence on this point has not yet led to a satisfactory result. Incidentally, this difficulty is not peculiar to my conceptual notation but recurs in a similar way in Peano's.

I had entertained the idea of a conceptual notation long before it assumed a more definite shape. The need to exclude with certainty any tacit presuppositions in the foundations of mathematics led me to the conceptual notation of 1879. My concern with the latter compelled me in turn to give a more exact formulation of the fundamental concepts of arithmetic, although I can no longer document this in detail. The recognition that the bearer of a number is not a heap, aggregate, or system of things but a concept may well have been furthered substantially by the conceptual notation. Instead of a concept one can also take its extension or, as I also put it, the relevant class. This distinction between a heap (aggregate, system) and a class, which had not perhaps been drawn as sharply before me, I owe, I believe, to my conceptual notation, although you will not perhaps discover any trace of it in reading my little work. Some of the reasonings that occurred in writing it may well have left no trace in print.

Yours sincerely,
G. Frege

VIII/3 [xxi/3]   FREGE to JOURDAIN   21.3.1904

Dear Mr Jourdain,

Many thanks for the papers you sent me and for your letter.[1] The question calls for mature reflection, and I am still not in a position to devote myself to this matter in the way it deserves. I still have some scientific letters to reply to, among others one by Mr B. Russell which I have left unanswered for too long. But I will think about the content of your letter as soon as I have time and then write to you in more detail.

Yours sincerely,
G. Frege

[Marginal note:] I hope you received the papers I sent you.

VIII/4 [xxi/4]   JOURDAIN to FREGE   22.3.1904

Little Close, Yateley,
Hants.
March 22, '04

Dear Prof. Frege

Many thanks for your memoirs & card. I quite agree with your remarks on Hilbert's 'Grundlagen ...'. I shall hope to hear from you when you have time.

Y^{rs} sincerely
Ph. E. B. Jourdain

VIII/5 [xxi/5]   JOURDAIN to FREGE   28.1.1909

Broadwindsor Manor,
Beaminster,
Dorset.
Jan. 28 '09

Dear Dr Frege,

I have tried to get copies of the papers mentioned in your 'Grundgesetze' as giving fuller information on Sinn, Bedeutung, Begriff, Gegenstand, Function, ... but cannot. Further, your separate works 'Function u. Begriff', and 'Ueber die Zahlen des Herrn Schubert' seem out of print.

VIII/3. 1 This postcard is an answer to a letter which has been lost. Jourdain's notebook gives an indication of the date and contents of the letter: The entry for 'March 10. 1904' contains the remark: '(To Frege: short account of getting over Russell's & Burali-Forti's contradiction by limitation on conception of a class – so that mathematical conceptions can apply to it)'. It is not known what papers Jourdain sent to Frege, or Frege to Jourdain.

Have you, then, spare copies of these which you can let me have for a short time to read?

Also, can you spare me separate copies of your: 'Ueber Sinn u. Bedeutung', 'Begriff u. Gegenstand', 'Formale Theorien der Arith.', 'Kritische Beleuchtung ... über Schröder', 'Über die Begriffsschrift des Peano ...', and any other papers you may have published since 1902. I think I have all your other works, & I have been convinced (chiefly by Mr B. Russell) that it is your ideas which are, perhaps, of the greatest importance in the present state of discussion of the principles of mathematics.

As your 0, 1, ... of the 'Grundgesetze' are what G. Cantor calls *cardinal* numbers or Mächtigkeiten, do you contemplate introducing (in a 3rd vol. of 'Grundgesetze') *ordinal* numbers 0, 1, ...?

Yours sincerely
Philip E. B. Jourdain

VIII/6 [xxi/6]   JOURDAIN to FREGE   15.2.1909

Broadwindsor Manor,
Beaminster,
Dorset.
Feb. 15 '09

Dear Dr Frege,

You will see by the enclosed card that my booksellers report that your 'Ueber die Zahlen ...' is not to be had. If my booksellers are mistaken, will you ask Pohle to send me a copy? I will then send him the price.

Many thanks for the pamphlets you sent me, which are very interesting, especially that on Peano.

Please send me copies of any others you may publish.

Yours sincerely,
Philip E. B. Jourdain

VIII/7 [xxi/7]   JOURDAIN to FREGE   16.4.1910

Broadwindsor
16.IV.'10

Dear Dr Frege,

Would you read over and criticize a typewritten MS. of mine (to appear eventually in the Quart. J. of Math.) on the development of your work on logic & the principles of mathematics? If so, I will send it to you registered.

Yours sincerely
Philip E. B. Jourdain

VIII/8 [xxi/8]   JOURDAIN to FREGE   23.4.1910

Broadwindsor
23.IV.'10

Dear Dr Frege,

I am to-day sending you the MS. Please keep it as long as you like: I shall feel grateful for any notes on it.

I mean to write a fuller account of the *Grundgesetze*, which I have not dealt with as fully as I should, But I hope the rest is fairly complete (together with parts of the section on Peano, which I am sending to P., and will send on to you later).

Yrs sincerely
Ph. Jourdain

VIII/9 [xxi/9]   FREGE to JOURDAIN   undated

Dear Mr Jourdain,

I am very grateful to you for the detailed account you devote to my logical and mathematical theories, and I hope that I shall gain some readers for my writings through you. I shall be pleased if the remarks I am enclosing are of use to you. You may use them as you see fit.[1]

With best regards,

Yours,
G. Frege

VIII/10 [xxi/10]   JOURDAIN to FREGE   29.3.1913

The Lodge,
Girton,
Cambridge.
March 29, 1913

Dear Prof. Frege,

In your last letter to me you spoke about working at the theory of irrational numbers. Do you mean that you are writing a third volume of the Grundgesetze der Arithmetik? Wittgenstein and I were rather disturbed to think that you might be doing so, because the theory of irrational numbers

---

VIII/9. 1 Jourdain published most of Frege's remarks, in English translation, as footnotes to the first 23 sections of his essay 'The Development of the Theories of Mathematical Logic and the Principles of Mathematics', *Quarterly Journal of Pure and Applied Mathematics* 43 (1912), pp. 237–69. The first 23 sections of Jourdain's essay, which are entitled 'Gottlob Frege', are reprinted as an appendix to this book, together with Frege's remarks in Jourdain's translation.

– unless you have got quite a new theory of them — would seem to require that the contradiction has been previously avoided; and the part dealing with irrational numbers on the new basis has been splendidly worked out by Russell and Whitehead in their Principia Mathematica. I say that we were disturbed, because we felt that you could not possibly, at this time, now that the theories you were the first to develop have been worked out by others, write a third volume that would be anywhere near your first and second in originality and brilliancy. You must forgive me for speaking like this; it is only because I feel that your work on the foundations of logic is so wonderfully fine that I should regret to see its continuation into a domain that has been already cultivated by others.

Another point. I am most anxious that a full and adequate biography of you should be published. Might I suggest that, in your spare time, you should draw up an autobiography which should give some account of the development of your thoughts on logic and arithmetic. I am most vividly anxious to have more information about this, the more so as I think that practically the whole of the development of the ideas must have gone on without any external influence. Would it be possible for you to write the biography, let me translate it, and have it published in America by the firm managed by Dr Carus?[1] I feel sure that he would be glad to do so; and it would be very pleasant to me to know that I had a small part to do in paying honour to you.

Yours very sincerely
Philip E. B. Jourdain

VIII/11 [xxi/11]   JOURDAIN to FREGE   15.1.1914[1]

The Lodge, Girton.
Cambridge, Jan. 15, 1914

Dear Prof. Frege,

Would you be kind enough to give me permission to translate part of your 'Grundgesetze' for 'The Monist'. I was thinking of the more popular parts (Bd. I, S. vi–xxvi, 1–8, 51–52; Bd. II. S. 69–80). If you will give me your permission, Wittgenstein has kindly offered to revise the translation, & then I would send it on to you.

VIII/10. 1 It is not known whether Frege composed such an autobiography. Dr Paul Carus was general editor of the *Monist*.
VIII/11. 1 With Frege's permission, a translation of parts of GGA. I appeared in the *Monist* 25 (1915), pp. 484–94, 26 (1916), pp. 182–99, and 27 (1917), pp. 114–27, but Jourdain collaborated on the translation with Johann Stachelroth, and it is not known how much of the translation is due to Jourdain.

Also, will you tell me if, in your opinion, (1) your theory of functions 'erster bzw. zweiter Stufe' is not the same as Russell's theory of orders (cf. Principia Math. I. 54ff); (2) whether you now regard assertion ($\vdash$) as merely psychological; (3) whether, in view of what seems to be a fact, namely, that Russell has shown that propositions can be analyzed into a form which only assumes that a name has a 'Bedeutung', & not a 'Sinn', you would hold that 'Sinn' was merely a psychological property of a name.

Yours sincerely

Philip E. B. Jourdain

VIII/12 [xxi/12]   FREGE to JOURDAIN   undated[1]

Dear Mr Jourdain,

I am very glad to give you permission to translate parts of my *Basic Laws* for the *Monist*. From your letter it seems to me that Mr Wittgenstein is again in Cambridge.[2] I had lengthy conversations with him before Christmas, and I wanted to write him a letter about them in order to carry on the thread, but I did not know where he was. Unfortunately I do not understand the English language well enough to be able to say definitely that Russell's theory (*Principia Mathematica* I, 54ff) agrees with my theory of functions of the first, second, etc. levels. It does seem so. But I do not understand all of it. It is not quite clear to me what Russell intends with his designation $\Phi!\hat{x}$. I never know for sure whether he is speaking of a sign or of its content. Does 'function' mean a sign? I already wrote to you once why I wanted to see the expression 'variable' banned. One never knows exactly whether it is supposed to be a sign or the content of a sign. On p. 54 Russell writes with reference to the proposition '$\Phi!\hat{x}$ implies $\Phi!a$ with all possible values of Φ': 'This makes a statement about x, but does not attribute to x a *predicate* in the special sense just defined'. If I understand this proposition right, I would write it in my conceptual notation as

$$\begin{array}{l} \\ \neg\!\!\stackrel{\mathfrak{f}}{}\!\!\overline{}\!\!\begin{array}{l} \mathfrak{f}(a) \\ \mathfrak{f}(x) \end{array} \end{array}$$

and in my opinion one could also write it as '$x = a$'. Why should this not be a case like any other where a predicate is being attributed to x? Do we not have a first-order function also in this case? With regard to your second question I want to say the following. Judging (or recognizing as true) is certainly an inner mental process; but that something is true is independent of the recognizing subject; it is objective. If I assert something as true I do

---

VIII/12. 1 An earlier draft of VIII/13.

2 [This is false. Wittgenstein had not been in England since October 1913.]

not want to talk about myself, about a process in my mind. And in order to understand it one does not need to know who asserted it. Whoever understands a proposition uttered with assertoric force adds to it his recognition of the truth. If a proposition uttered with assertoric force expresses a false thought, then it is logically useless and cannot strictly speaking be understood. A proposition uttered without assertoric force can be logically useful even though it expresses a false thought, e.g., as part (antecedent) of another proposition. What is to serve as the premise of an inference must be true. Accordingly, in presenting an inference, one must utter the premises with assertoric force, for the truth of the premises is essential to the correctness of the inference. If in representing an inference in my conceptual notation one were to leave out the judgement strokes before the premised propositions, something essential would be missing. And it is good if this essential thing is visibly embodied in a sign and not just added to it in the act of understanding according to a tacit convention; for a convention according to which something has to be added in that act of understanding under certain circumstances is easily forgotten even if it was once stated explicitly. And so it happens that something essential is completely overlooked because it has not found an embodiment. But what is essential to an inference must be counted as part of logic.

As far as your third question is concerned, I do not believe that we can dispense with the sense of a name in logic; for a proposition must have a sense if it is to be useful. But a proposition consists of parts which must somehow contribute to the expression of the sense of the proposition, so they themselves must somehow have a sense. Take the proposition 'Etna is higher than Vesuvius'. This contains the name 'Etna', which occurs also in other propositions, e.g., in the proposition 'Etna is in Sicily'. The possibility of our understanding propositions which we have never heard before rests evidently on this, that we construct the sense of a proposition out of parts that correspond to the words. If we find the same word in two propositions, e.g., 'Etna', then we also recognize something common to the corresponding thoughts, something corresponding to this word. Without this, language in the proper sense would be impossible. We could indeed adopt the convention that certain signs were to express certain thoughts, like railway signals ('The track is clear'); but in this way we would always be restricted to a very narrow area, and we could not form a completely new proposition, one which would be understood by another person even though no special convention had been adopted beforehand for this case. Now that part of the thought which corresponds to the name 'Etna' cannot be Mount Etna itself; it cannot be the meaning of this name. For each individual piece of frozen, solidified lava which is part of Mount Etna would then also be part of the thought that Etna is higher than Vesuvius. But it seems to me absurd that pieces of lava, even pieces of which I had no knowledge, should be parts of my thought. Thus both things seem to me necessary: (1) the meaning of a name, which is that about which something is being said, and (2) the sense of the name, which is part of the thought. Without meaning, we could indeed have a thought, but only a mythological or literary thought, not a thought that could further scientific knowledge. Without a sense, we would have no thought, and hence also nothing that we could recognize as true.

To this can be added the following. Let us suppose an explorer travelling in an unexplored country sees a high snow-capped mountain on the northern horizon. By making inquiries among the natives he learns that its name is 'Aphla'. By sighting it from different points he determines its position as exactly as possible, enters it in a map, and writes in his diary: 'Aphla is at least 5000 metres high'. Another explorer sees a snow-capped mountain on the southern horizon and learns that it is called Ateb. He enters it in his map under this name. Later comparison shows that both explorers saw the same mountain. Now the content of the proposition 'Ateb is Aphla' is far from being a mere consequence of the principle of identity, but contains a valuable piece of geographical knowledge. What is stated in the proposition 'Ateb is Aphla' is certainly not the same thing as the content of the proposition 'Ateb is Ateb'. Now if what corresponded to the name 'Aphla' as part of the thought was the meaning of the name and hence the mountain itself, then this would be the same in both thoughts. The thought expressed in the proposition 'Ateb is Aphla' would have to coincide with the one in 'Ateb is Ateb', which is far from being the case. What corresponds to the name 'Ateb' as part of the thought must therefore be different from what corresponds to the name 'Aphla' as part of the thought. This cannot therefore be the meaning which is the same for both names, but must be something which is different in the two cases, and I say accordingly that the sense of the name 'Ateb' is different from the sense of the name 'Aphla'. Accordingly, the sense of the proposition 'Ateb is at least 5000 metres high' is also different from the sense of the proposition 'Aphla is at least 5000 metres high'. Someone who takes the latter to be true need not therefore take the former to be true. An object can be determined in different ways, and every one of these ways of determining it can give rise to a special name, and these different names then have different senses; for it is not self-evident that it is the same object which is being determined in different ways. We find this in astronomy in the case of planetoids and comets. Now if the sense of a name was something subjective, then the sense of the proposition in which the name occurs, and hence the thought, would also be something subjective, and the thought one man connects with this proposition would be different from the thought another man connects with it; a common store of thoughts, a common science would be impossible. It would be impossible for something one man said to contradict what another man said, because the two would not express the same thought at all, but each his own.

For these reasons I believe that the sense of a name is not something subjective [crossed out: in one's mental life], that it does not therefore belong to psychology, and that it is indispensable.

VIII/13 [xxi/13]   FREGE to JOURDAIN   28.1.1914

Brunshaupten
28 January 1914

Dear Mr Jourdain,

I am very glad to give you permission to translate part of my *Grundgesetze* for the *Monist*.

From your letter it seems to me that Wittgenstein is back in Cambridge. I had lengthy conversations with him before Christmas, and I would like to write to him so that something fruitful will come out of them, but I do not know where he is. Unfortunately, I do not understand Russell's *Principia* well enough to be able to say with certainty that Russell's theory (*Principia* I, 54ff) agrees with my theory of first- and second-level functions. I already wrote to you once why I should like to ban the expression 'variable'. I find it very difficult to read Russell's *Principia*; I stumble over almost every sentence. Thus I read on p. 4:

'In mathematical logic, any symbol whose meaning is not determined is called a *variable*, and the various determinations of which its meaning is susceptible are called the *values* of the variable.'

Thus according to Russell a symbol has a meaning (*Bedeutung*). There are symbols which have a meaning which is not determined. Russell wants to call such signs *variables*. One variable can differ from another in that its meaning admits (is 'susceptible' of) different determinations from the other. Neither the former nor the latter meaning is determined; each of them admits several determinations; but the determinations which the former meaning admits are different from the determinations which the latter meaning admits. To me this is quite mysterious. There are no undetermined men. I also do not believe that there are meanings with such peculiar properties as the ones I just gave following Russell. Instead of saying that 'the meaning of this sign is not determined', one should say that 'it is not determined what meaning this sign is to have'. Before it is established what meaning a sign is to have, one must not use the expression 'the meaning of this sign', and one must say neither 'the meaning of this sign is determined' nor 'the meaning of this sign is not determined.'

On p. 5 Russell says, 'Variables will be denoted by single letters, and so will certain constants.'

If a variable is designated [or denoted] by a single letter, then the meaning of this letter is a variable, and hence a symbol according to Russell, and the meaning of this symbol is not determined. It must then be a sign of a sign. It seems that constants, too, are supposed to be symbols, but symbols whose meanings are determined, and some of these symbols, too, are supposed to be designated by individual letters.

There further occurs the expression 'variable propositions'. Accordingly it is to be assumed that a proposition according to Russell is a symbol. This seems to decide the question whether a proposition for Russell is a proposition, a sign composed of signs, or whether a proposition is the sense of a proposition and hence what I call a 'thought'. The letter '*p*' will accordingly designate a symbol and thus be a sign of a sign.

There also occurs the expression 'variable functions'. Accordingly a function, at least a variable one, will be a symbol. But a constant function is probably also a symbol, though one whose meaning is determined. The letter '*f*' is then a sign of a sign.

This interpretation is confirmed by the sentence on p. 6: 'An aggregation of propositions, considered as wholes not necessarily unambiguously determined into a single proposition ... is a function *with propositions as arguments*'. According to this, a proposition composed of propositions (aggregation of propositions into a single proposition) is in some circumstances a function. Now since a composite proposition is a symbol, a composite sign, a function with propositions as arguments is also a symbol.

There now occurs the proposition '*p* is true'. Now the letter '*p*' designates a symbol; hence *p* is a symbol. Now can a symbol be true? Surely only if it has a sense. And is it not then the sense that is true, properly speaking? On p. 8 I find the sentence '*A* believes *p*'. Can one believe a symbol, a sign?

On p. 15 it says: '*Propositional functions*. Let φx be a statement containing a variable *x* and such that it becomes a proposition when *x* is given any fixed determined meaning'.

Since a variable is a symbol, it is to be assumed that a statement is a composite sign and that as one of its parts it contains a symbol which is a variable. But the group of letters 'φx' is not itself this composite sign, nor is the letter '*x*' this variable, but this variable is a symbol which is not visible at all in Russell's text. We also do not learn what it looks like. It is designated by the letter '*x*'. This symbol *x* is thus unknown to us. Russell only tells us that it has a meaning (*Bedeutung*), but not a determined one. But this symbol can be given a fixed determined meaning according to Russell. But is this admissible? For the unknown symbol *x* (not the letter '*x*') already has a meaning, viz. an undetermined one, and can we then give it a second meaning, namely a determined one? Or did I misunderstand Russell's words? Perhaps the sentence 'Variables will be denoted by single letters' does not mean: Variables will be designated by individual letters, but: Variables are individual letters? Then an individual letter, e.g. '*x*', would itself be the symbol whose meaning was undetermined. It would then be better to write Russell's sentence above as: 'Let "φx" be a statement containing a variable "*x*" and such that it becomes a proposition when "*x*" is given any fixed determined meaning'.[1] In this case, too, a statement is a composite sign; but

---

VIII/13. 1 Here follows a crossed-out sentence: 'The quotation marks are then supposed to indicate that something is being said about the group of letters "φx" itself, or about the letter "*x*" itself, and not about its meaning.'

as one of its parts it now contains the letter 'x' instead of the unknown symbol x. Russell's sentence 'and so will certain constants' should then be paraphrased as: 'certain constants are letters'. Then the group of letters 'φx' would not mean anything, but would stand for a group of signs containing 'x', e.g., the group 'x is hurt', and this would be a statement. This becomes a proposition if we write (say) 'Alfred' for 'x', or more generally, if we replace the letter 'x' by a name having a meaning. This again shows that a proposition is a (composite) sign. But what, then, is a propositional function? Russell writes: 'Then φx is called a "propositional function".' This is evidently supposed to be a definition. Let us look at the composite sign 'x is hurt', and let us suppose first that the letter 'x' is not a variable but that it designates one. Let the variable be the symbol '⋈'. Then the letter 'x' designates the symbol '⋈'. Instead of this we can also say: then x is the symbol '⋈'. Wherever the letter 'x' occurs, we can then insert 'the symbol "⋈"'. Thus instead of 'x is a variable' we shall say 'the symbol "⋈" is a variable', and instead of 'x is hurt' we shall say 'the symbol "⋈" is hurt' or also 'the variable "⋈" is hurt'. This will not of course be intelligible without further explanation; but we can also imagine the words 'is hurt' explained in such a way that 'the symbol "⋈" is hurt' is intelligible; or better: instead of 'is hurt' we write 'is black'. Given our supposition that the letter 'x' designates the symbol '⋈', we can then write 'the symbol "⋈" is black' or 'the variable "⋈" is black' instead of 'x is black'. Now we need not know at all what the meaning of the symbol '⋈' is; for we are not saying anything about this meaning, but about the symbol '⋈' itself. Even if this symbol has no meaning at all, this would not need to bother us. Then 'x is black' would be a proposition (*Satz*). And this proposition would contain a statement about an object which was designated by the letter 'x'. But then we cannot say that 'x is black' is a propositional function; for we can no longer give 'x' a meaning arbitrarily because it already has a meaning; namely, it designates the variable '⋈'. It is almost certain that this is not what Russell means. We must then suppose that the letter 'x' does not designate a variable but that it is one, and that it is a symbol whose meaning according to Russell is not determined. Now if the letter 'x' had a meaning, even if only an undetermined one, then 'x is hurt' would contain an assertion about this meaning. But since there are no undetermined meanings, the letter 'x' has no meaning at all if it is supposed to be (and not to mean) a variable. And 'x is hurt' does not contain an assertion about some object or person. This agrees with Russell's sentence: 'Thus "x is hurt" makes no assertion at all, till we have settled who x is'. The subsidiary clause could be paraphrased as: 'till we have given the letter "x" a meaning'. Accordingly, Russell's meaning seems to be that the composite sign 'x is hurt' is a statement which contains the variable 'x', and that this variable 'x' is the letter 'x' itself. I believe that Russell's sentence 'Then φx is called a "propositional function"' must be understood as: 'Then "φx" is called a "propositional function"'. What he called a 'statement' he now calls a 'propositional function'. Accordingly, the composite sign 'x is hurt' is according to Russell a propositional function. Now it has not been laid down what meaning (*Bedeutung*) this composite sign 'x is hurt' is supposed to have, and it is accordingly a variable. We shall be able to say: according to Russell every propositional function is a variable. But this does not agree with Russell's statement 'Thus "x is hurt" is an ambiguous "value" of a propositional function'. What seemed to be a propositional function is now an ambiguous value of a propositional function.[2] Russell writes: 'When we wish to speak of the propositional function corresponding to "x is hurt", we shall write "$\hat{x}$ is hurt"'. Accordingly the composite sign '$\hat{x}$ is hurt' seems to be the name of a *propositional function*. But this only gives us the name and not what is supposed to be designated by this name.[3]

Incidentally, the expression 'propositional function' occurs already on p. 6 without, as far as I can see, having been defined previously.

Now we must still look at the expression 'ambiguous value'. Here we must remember that according to Russell a propositional function is a variable and that 'the various determinations of which its meaning is susceptible are called the values of the variable'. Accordingly we must suppose that the value of a propositional function is a determination of its meaning. But how can the value then be ambiguous? What is ambiguous? The inkpot that stands in front of me is not ambiguous, no more is my penholder. Only a sign (word) can be ambiguous, something that has the purpose of designating something. But if a sign is ambiguous, then this is a mistake.

Accordingly, if a value is ambiguous, then it must be a sign, but a sign that must really not be used just because of its ambiguity. Indeed, Russell encloses 'x is hurt' in quotation marks, which seems to indicate that he means the sign itself and not some content designated by it. And yet one would think that a value of a function cannot be a mere sign.

We do at least agree that when we use a sign (symbol, word) in the context of a proposition we speak of what this sign designates. If I want to say something in writing about a written sign, I enclose it in quotation marks, and the resulting composite sign is then the name of a sign. On the whole, Russell seems to follow the same usage; but sometimes he may be departing from it.

It seems to me that the difficulties keep piling up as one penetrates further into Russell's work. I will therefore suspend any further study of it till my doubts have been resolved. If you could help me to a better understanding of Russell's real meaning, I shall be grateful to you. In the meantime I remain

Yours sincerely,
G. Frege

---

2 Here follow 6 crossed-out lines in the original.
3 Here follow 53 crossed-out lines in the original.

IX FREGE–KORSELT

Editor's Introduction

Alwin Reinhold Korselt (1864–1947) taught at the secondary school of Plauen in Saxony. His most outstanding contribution to the algebra of logic was the proposition that not every formula of first-order two-place quantificational logic is expressible ('condensible') in Schröder's relational calculus. This proposition was first published by Leopold Löwenheim on the basis of a written communication from Korselt.[1] It was later sharpened by Tarski (1941) and Lyndon (1950), which is why it now bears the designation 'Korselt–Tarski proposition'. In an essay 'Über mathematische Erkenntnis',[2] Korselt defended Frege's definition of the predicate 'number' as well as of the specific numbers 'one' and 'endless'.[3] And in the discussion he carried on with Frege from 1903 to 1908 about the foundations of geometry, he tried to adopt a position intermediate between Frege's and Hilbert's views.[4] The correspondence is concerned with Korselt's proposed solution to Russell's paradox. In a letter dated 26.6.1903, which is now lost, Frege rejected the solution Korselt proposed in his first letter. A discussion of the antinomies diverging from the solutions proposed in these letters is to be found in Korselt's essays 'Paradoxien der Mengenlehre'[5] and 'Auflösung einiger Paradoxien'.[6] His most important contributions to the algebra of logic and the foundations of mathematics are listed in Alonzo Church, 'A Bibliography of Symbolic Logic'.[7]

IX/1 [xxv/1]   KORSELT to FREGE   21.6.1903

Plauen i.V.

21 June 1903

Dear Sir,

Since I have been concerned with the algebra of logic for years and know your works, allow me to make a remark on the contradiction you mention in your book, *The Basic Laws of Arithmetic*, vol. II, p. 254. If a contradiction appears, this is only a sign that the account is based on a definition which cannot for some reason be used to give a non-contradictory account – whether or not the reason can be exactly designated – and that the definition must be replaced by another which fulfils the desired purpose.

---

1 Cf. L. Löwenheim, 'Über Möglichkeiten im Relativkalkül', *Mathematische Annalen* 76 (1915), pp. 447–70, sect. 2.
2 *Jahresberichte der Deutschen Mathematiker-Vereinigung* 20 (1911), pp. 364–80, esp. p. 371.
3 Cf. PW, p. 214, note 2.
4 Cf. Frege, GLG I, II, III 1, III 2, III 3, and Korselt, 'Über die Grundlagen der Geometrie', *Jahresberichte der Deutschen Mathematiker-Vereinigung* 12 (1903), pp. 402–7, and 'Über die Logik der Geometrie', ibid. 17 (1908), pp. 98–124.
5 ibid. 15 (1906), pp. 215–19.
6 ibid. 25 (1916–17), pp. 132–8.
7 *Journal of Symbolic Logic* 1 (1936), p. 145.

This new definition will be:

Something, called *s*, is a member of a class, called *e* ≡ what is designated by *s is not a class* and falls under the concept whose extension is designated by *e*.

Thus if your class *c* is a member of itself, then it is *not* a class according to *my* definition – which is a contradiction.

But that your class *c* is *not* a member of itself is according to my definition the same as the assertion:

'The (composite) assertion that *c* is not a class and falls under the concept whose extension it is, is false',

and the assertion enclosed in quotes is true.

With this any contradiction has disappeared, and it follows generally:

No class is a member of itself.

The law of the excluded middle holds also for 'membership of a class' because of the two opposed assertions:

a class *c* is a member of itself,
a class *c* is not a member of itself,

the first is always false and the second always true. All we need to do is hold on to the fact that a class is not a concept but the object of a concept and that a concept is not a class, a fact which you demonstrated satisfactorily in your essay 'On Concept and Object'.

I agree on most points with your views on 'formal arithmetic',[1] but on the basis of my own investigations I believe that in some cases your criticism is only valid for its currently familiar state, or that you do not grant the mathematicians you criticize the same freedom of expression that you must demand for yourself.

> Yours sincerely,
> Dr phil. Alwin Korselt

IX/2 [xxv/3]   KORSELT to FREGE   27.6.1903

> Plauen i.V.
> 27 June 1903

Dear Sir,

My definition is only the beginning of a series of definitions for all classes expressible in language, namely:

(1) Individual, individual object = first-level object = zero-level class (Socrates, *individuum signatum*).

IX/1 1 The reference is probably to GGA II, sects 86–187.

(2) Class of first-level objects = second-level object = first-level class, whose basis is a first-level concept (sage, class of satellites of the earth).

－－－－－

(n) Class of nth-level objects $= n + 1^{th}$-level object (whose basis is an nth-level concept).

A class a is a member of class b if a is one level lower than b and if a falls under the 'basis of b'.

Then a class can be a member of another class but never of itself. Your doubt has been removed. '*The* moon' is a member of the '*class* of moons', but this *class* is not a member of itself, even though it consists of only one individual. Thus it will have to be laid down that the extension of a concept does not fall under the concept itself, even if the concept happens to be 'one-numbered'. What has been designated as a 'class' must not be equated with an individual. This linguistic rigidity is evidently as unavoidable as the following:

The concept horse is not a concept.[1]

It must surely be possible to dispose of Russell's 'sophism' as well as Schröder's![2]

The question what is to be regarded as a class or as an individual can be given different answers in different investigations; one must stick to the same sense only in one and the same investigation.

It can be demonstrated by using simple examples that Cantor's propositions A and D (*Annalen* 46, sect. 2)[3] and Schröder's proposition:

IX/2. 1 The reference is evidently to *BuG*, p. 196.

2 The reference is presumably to ⟨13⟩ where Frege attacks Schröder's identification of an individual with the class containing this and only this individual as an element and brings out the following contradiction within it. If P is a class which contains (among others) the two different elements a and b, and Q is a class which contains class P as its only element, and if – according to Schröder's identification – P is identical with the class $\{P\}$ which contains P and only P as element, then $P = \{P\} = Q$, and hence a and b are also elements of Q; but since Q contains only P as element, a and b must be identical with P and hence also with each other, which contradicts the supposition that they are different. Frege refers to this result in ⟨13⟩ as a 'sophism'. It is not known whether he himself referred to this sophism in his previous letter (now lost).

3 Cf. Georg Cantor, 'Beiträge zur Begründung der transfiniten Mengenlehre' (first article), *Mathematische Annalen* 46 (1895), pp. 481–512. Propositions A and D read as follows: 'If a and b are any two cardinal numbers, then either $a = b$ or $a < b$ or $a > b$' (A) and 'If two sets M and N satisfy the condition that N is not identical either with M itself or with part of M, then there is a part N_1 of N which is equivalent to M' (D). These propositions correspond to propositions (27) A_0 and (27) D in Ernst Schröder, 'Ueber zwei Definitionen der Endlichkeit und G. Cantor'sche Sätze', *Nova Acta: Abhandlungen der Kaiserlichen Leopold-Carolinischen Deutschen Akademie der Naturforscher*, vol. 71, no. 6 (1898), pp. 303–62, esp. p. 317. Korselt's discussion refers to the latter source.

Every set *can* be simply ordered ('On two Definitions of Finitude and G. Cantor's Propositions')[4] cannot be derived from the presently known propositions of the theory of manifolds and exact logic. The 'can' does not even have a psychological value.

Schröder's proof of proposition 24 contains an incurable gap, as Schröder himself recognized,[5] but can be transformed into a strict proof by a sharper use of the 'one-one reversible projection', and since this proof does not make use of 'complete induction', it is conceptually simpler than Bernstein's proof.[6]

By making use of that projection I proved proposition 21 (loc. cit., p. 317),[7] thus answering Schröder's 'question' on p. 326 in the affirmative.[8]

With many thanks for the reference to Russell's new work,

Yours sincerely,
Dr A. Korselt

IX/3 [xxv/4]   KORSELT to FREGE   30.6.1903

Plauen i.V.
30 June 1903

Dear Sir,

There is an even simpler solution to Russell's sophism than the one I last indicated; which goes as follows:

Neither mankind nor the negro race (the totality of negroes) nor the citizenry of Plauen falls under the concept man; instead the following principle holds for every concept, requiring it to be not only clear but also distinct: *If DIFFERENT objects of concept a fall under the extension of concept b, the extension of concept b cannot fall under concept a (P).*

While this principle has been used (discovered?) by Schröder, it has

---

4 Cf. Schröder, loc. cit., p. 356: 'As our next fundamental proposition we might put forward the following: *Theorem*. Any set *can* be simply ordered.'

5 In Schröder, loc. cit., p. 317, the proposition reads:

$$(a \in b \sim c \in a) \in (b \sim a \sim c)$$

which means: If a superset b of a set a is equipollent to a subset c of a, then the sets a, b and c are equipollent to one another.

6 First published in E. Borel, *Leçons sur la théorie des fonctions* (Paris 1898), pp. 103ff.

7 The proposition reads:

$$(a \sim d \in b) \in \sum_c (a \in c \sim b)$$

which means: If a set a is equipollent to a subset d of set b, then there is a superset c of a equipollent to b. For Schröder's formulation, cf. loc. cit., p. 319.

8 Cf. Schröder, loc. cit., p. 326: 'It is still an open question whether our proposition (21) has absolute validity in a realm of thought which is conceptually completely determined, that is, which, though infinite, is nevertheless complete in itself...'

never been translated into a formal language. According to *P* and your definition of 'membership', no class can be a member of itself. While it follows from this that the 'class of classes that are not members of themselves' does *not* fall under the concept *class which is not a member of itself* (*c*), it does not follow from this that it falls under the concept *class which is a member of itself*, but that it can contain, as it actually does, *different* objects of concept *c*, and this is no longer a contradiction.

Yours sincerely,
Dr A. Korselt

X FREGE–LIEBMANN

Editor's Introduction

Heinrich Liebmann (1874–1939) was professor of mathematics at Heidelberg from 1920 to 1935. From 1897 to 1899 he was an assistant at the Mathematical Model Collection in Göttingen. In the winter semester 1898–9 he heard Hilbert's lecture, *Elemente der Euklidischen Geometrie*, which was edited by Hilbert's assistant, H. von Schaper, and issued in 70 autographed copies, and which was the forerunner of Hilbert's *Grundlagen der Geometrie*. H. Liebmann passed an edited transcript on to Frege, whom he knew as a colleague of his father, Otto Liebmann, professor of philosophy at Jena. This was the occasion for two letters from Frege which, as far as is known, was the extent of their correspondence.

X/1 [xxvii/1] FREGE to LIEBMANN 29.7.1900

Jena

29 July 1900

My dear Doctor,

I am returning herewith the copy of Hilbert's lecture with many thanks. Clever and inventive as it is in many points, I think that it is on the whole a failure and in any case that it can be used only after thorough criticism. Hilbert, like many mathematicians, seems to lack a clear awareness of what a definition can do and that it cannot do the same thing as a theorem, so that a definition cannot take the place of a theorem. He uses first- and second-level characteristic marks – according to my distinction – side by side to define the same concept, as if one were to say:

'The concept God is given by the following axioms:

      (a) a God is omnipotent;

      (b) a God is omniscient;

      (c) there is a God.'[1]

The last one is second-level, and must not therefore be aligned with the first two characteristic marks, which are first-level. Instead, after defining the concept by means of the first two characteristic marks, one should then prove that there was something of the kind; and one must not try to by-pass such a proof by means of a definition. As I learned from a letter from Prof. Hilbert himself, his axioms are to be regarded as parts of his definitions,[2] so that, e.g., the first chapter of his *Foundations of Geometry*, with all the axioms and propositions it contains, is but a single definition. The axioms

---

X/1. 1 Cf. IV/5 above.

2 Cf. IV/4 above.

are thus supposed to give the individual determinations of a concept. But here is still another monstrosity: it is not a single concept but three (point, line, plane) that are supposed to be defined at the same time in one definition extending almost over one printer's sheet. Hilbert completely blurs the distinction between first- and second-level concepts, so that one can hardly ever tell for sure whether he wants to define a first- or second-level concept. As far as the lack of contradiction and mutual independence of the axioms is concerned, Hilbert's investigation of these questions is vitiated by the fact that the sense of the axioms is by no means securely fixed; they are supposed to help define, e.g., the concept of a line, and yet the word 'line' itself occurs in them, and not just 'line' but also 'point' and 'plane', which are themselves still to be explained. This is like Münchhausen pulling himself out of a swamp by his own topknot. The independence of an axiom A from others is the lack of contradiction between the contradictory of A and the other axioms. I have reasons for believing that the mutual independence of the axioms of *Euclidean* geometry cannot be proved. Hilbert tries to do it by widening the area so that Euclidean geometry appears as a special case;[3] and in this wider area he can now show lack of contradiction by examples; but only in this wider area; for from lack of contradiction in a more comprehensive area we cannot infer lack of contradiction in a narrower area; for contradictions might enter in just because of the restriction. The converse inference is of course permissible. Hilbert was apparently deceived by the wording. If an axiom is worded in the same way, it is very easy to believe that it is the same axiom. But it depends on the sense; and this is different, depending on whether the words 'point', 'line', etc. are understood in the sense of Euclidean geometry or in a wider sense. I have communicated these and other doubts to Hilbert in a letter, but have not received an answer. But according to what he told Prof. Gutzmer,[4] he seems no longer quite sure of his proofs, at least partly because of my doubts, and is devising new ones. If he succeeds, it may be of great importance for mathematics; but I still have strong doubts about it. If you do not think my doubts justified, I should be interested to learn your reasons.

    With best regards,

Yours sincerely,

G. Frege

---

3 Hilbert proves that his system of geometrical axioms is free from contradiction by giving an 'analytical' model within the area of certain algebraic numbers (*Grundlagen der Geometrie*, 1st ed. (1899), sect. 9). In such a proof the axioms are conceived, not as geometrical statements, but as statement *forms* which admit different (analytical, geometrical, etc.) interpretations, and this leads Frege to speak, somewhat misleadingly, of 'widening the area' of the axioms.

4 C. F. A. Gutzmer was at the time professor of mathematics at Jena.

X/2 [xxvii/2]   FREGE to LIEBMANN   25.8.1900

Bad Steben
25 August 1900

Dear Doctor,

Since you are interested in the foundations of geometry, I am sending you
a copy of my correspondence with Prof. Hilbert about his *Festschrift*.[1] As
you can tell from this, it is of some concern to me that you gain a more
precise knowledge of my views on this matter. The correspondence has,
however, been at a standstill for more than half a year; but I have a postcard
from Prof. Hilbert announcing a continuation of it.[2] Perhaps there is already
a letter waiting for me in Jena and my wife is keeping it from me with the
best of intentions.[3] In re-reading my letters I got the impression that I judged
Hilbert's investigations more favourably at the time than I do now. But I
must await Prof. Hilbert's answer. I have proposed to him that we might
envisage an eventual publication of our correspondence because of the
importance of the questions discussed in it.[4]

I will now try to explain to you in brief what I understand by second-level
concepts.

I must first emphasize the radical difference between concepts and
objects, which is such that a concept can never stand for an object or an
object for a concept. It is impossible to give real definitions here. The essence
of concepts can be characterized by saying that they have a predicative
nature. An object can never be predicated of anything. When I say 'the
evening star is Venus', I do not predicate Venus but *coinciding with Venus*.
In language, proper names correspond to objects, concept words (*nomina
appellativa*) to concepts. Yet in language the sharpness of the difference is
somewhat blurred, in that what were originally proper names (e.g. 'moon')
can become concept words, and what were originally concept words (e.g.
'God') can become proper names. Concept words come with the indefinite
article and with words like 'all', 'some', 'many', etc. Here there occur many
subtleties which I will not go into now. Between objects and (first-level)
concepts there occurs now the relation of subsumption: an object falls under
a concept; e.g., Jena is a university town. Concepts are usually composed of
partial concepts, the characteristic marks. *Black silk cloth* has the
characteristic marks *black*, *silk* and *cloth*. An object which falls under this
concept has these characteristic marks as its properties. What is a
characteristic mark with respect to a concept is a *property* of an object

---

X/2. 1 The copy referred to was a copy of the important parts of IV/3 to IV/5
above.

2 Cf. IV/6 above.

3 Frege never received a detailed reply to IV/5 above.

4 The publication proposed by Frege never materialized because of Hilbert's
reluctance. Cf. Editor's Introduction to IV above.

falling under that concept. Very different from subsumption is the subordination of a first-level concept under a first-level concept: 'All squares are rectangles'. The *characteristic marks* of the superordinate concept (rectangle) are also *characteristic marks* of the subordinate one (square). When I say, 'There is at least one square root of 4', I do not predicate anything of 2 or −2, but of the concept square root of 4. Nor do I give a characteristic mark of this concept, for this concept must already be completely known. I do not pick out a component part of the concept, but give a certain character of it whereby it differs, e.g., from the concept *even prime number greater than 2*. I compare the individual characteristic marks of a concept with the stones that constitute a house, and what is predicated in our proposition to a property of the house, e.g., its spaciousness. Here, too, something is predicated, but it is not a first-level but a second-level concept. The way Jena is related to *university town* is quite similar to the way *square root of 4* is related to there-is existence. We have here a relation between concepts, though not between first-level concepts as in subordination, but between a first-level concept and a second-level concept, which is similar to the subsumption of an object under a first-level concept. The first-level concept here plays a part similar to that of the object in subsumption, and the second-level concept a part similar to that of the first-level concept there. We could also speak of subsumption here; however, the relation here, though similar, is not the same as the subsumption of an object under a first-level concept. I want to say that a first-level concept falls (not under but) into a second-level concept. The difference between first- and second-level concepts is just as sharp as the one between objects and first-level concepts; for objects can never stand for concepts; hence an object can never fall under a second-level concept; this would not be false but nonsensical. If one were to try some such thing in language, one would get neither a true thought nor a false one, but no thought at all. Incidentally, I once published something on concept and object in Avenarius's journal.[5] Another character of a first-level concept is given in the proposition that if an object falls under this concept, then an object different from this one also falls under this concept. Here we have a second second-level concept. Out of these two concepts, considered as second-level characteristic marks, we can form a third second-level concept into which fall all first-level concepts under which fall at least two different objects. The concepts *prime number*, *planet*, *man* would be concepts falling into this second-level concept. It seems to me that Prof. Hilbert first had the idea in mind of defining second-level concepts; but he does not distinguish them from first-level concepts.[6] And this explains what Hilbert's explanations always leave unclear: how the same concept seems to be defined twice. It just is not the same concept. At first it is a second-level concept, afterwards a first-level concept falling into

---

5 The reference is to *BuG*.
6 Cf. IV/5 above.

it. It is a mistake to mix up the two and to use the same word (e.g. 'point') in both connections.

With warm regards,

Yours sincerely,
G. Frege

XI FREGE–LINKE

Editor's Introduction

Paul F. Linke (1876–1955) was a pupil of Lipps, Wundt and Bucken, and strongly influenced by Brentano, Husserl and Scheler. In 1907 he qualified as a university lecturer at Jena, where he remained for the rest of his life. Frege and Linke knew each other very well personally. Linke was also one of the first philosophers to refer to Frege in his publications. Linke later published a monograph on Frege. 'Gottlob Frege als Philosoph', *Zeitschrift für philosophische Forschung* 1 (1946–7). A typewritten copy of the following letter from Frege contains a number of notes in Linke's hand which have been added as footnotes to the letter.

XI/1 [xxviii/2]   FREGE to LINKE   24.8.1919

Bad Kleinen in Mecklenburg
24 August 1919

Dear Colleague,

Many thanks for sending me your essay and for the way you mention me in it.[1] I am in complete agreement with what you say about literati and relativism. 'Formal' is one of those words that one likes to use where one lacks a concept. I am indeed doubtful whether you have succeeded in laying the foundations of ethics, though this would be extremely desirable. It seems to me that this question must be reserved for further reflection.

I found a letter from you dated 2 August 1916, which was to serve as an introduction to a verbal discussion. If I am not mistaken, you also came to see me with our colleague Koebe;[2] but as far as I remember there was hardly any talk of the content of your letter. In it you dealt with the question whether the mathematical equals sign means equality or identity. You wrote at the time: 'Equality is a special case of difference and means identity of several different objects in a certain respect or with reference to an (ideal) characteristic mark'.

I should speak of 'agreement' instead of 'identity' here;[3] for properly speaking different objects cannot be identical at all, though they can agree in

---

XI/1. 1 The reference is to Linke's essay, 'König Literat und die Ethik', *Die Tat: Monatsschrift für die Zukunft deutscher Kultur*, Jena 11 (1919–20), pp. 359–81. On pp. 362ff Linke appeals to Frege's GGA I in attacking stipulative definitions and the conventionalism that goes with them.

2 Paul Koebe, a professor of mathematics, transferred from Leipzig to Jena in 1914. The letter referred to was in Scholz's possession but is now lost.

3 Linke notes:

'To *agree* in part or specifically in a character (= dependent part)' means: to have this part or this character (this 'characteristic mark') in common: two houses have in common a certain wall, their owner, their height, their purchase price = they have the *same* wall, the *same* owner, the *same* height, the *same* purchase price. L.

some respect, e.g., in colour. If equality is not identity but agreement in some respect, then the word 'equal' is almost without content, unless one adds to it in what respect the agreement is meant; for if two objects are given, it will almost always be possible to specify some respect in which they agree. Most mathematicians probably think that the equals sign does not mean identity. But asked what it means, or in what respect what is designated on the left- and right-hand sides agrees, they would probably give very different answers.

We do not quite agree in the use of the words 'characteristic mark' and 'property', 'concept' and 'object'. I distinguish the first two as follows: Blue, silk, and ribbon are characteristic marks of the concept blue silk ribbon, not its properties, because the concept is neither blue nor silk nor a ribbon. But an individual thing which I hold in my hand may fall under the concept blue silk ribbon, in which case it is blue.[4] Blue is then one of its properties. A characteristic mark of a concept is a property of an object falling under it. The predicative use is characteristic of a sign meaning a concept. Properly speaking, the copula must be counted as part of the sign of the concept. Apart from the copula, such a sign may consist of an adjective or of a *nomen appellativum* which may be accompanied by attributes and which appears either without article or with the indefinite article. Instead of the copula combined with an adjective or with a *nomen appellativum*, there can also appear a verb in the third person. An object can be designated either by a proper name or by a sign – e.g., 'the victor of Austerlitz' – which can stand for a proper name (individual name). An individual name is formed out of a *nomen appellativum* by adding the definite article or a demonstrative pronoun to it. But properly speaking this is permissible only if the concept of the *nomen appellativum* is not empty and if only one object falls under it. When an individual name is formed in this way, the object designated by it falls under the concept of the *nomen appellativum* and has the characteristic marks of this concept as its properties.*

As an example of identity you cite in your letter: 'The object meant by the concepts loser of Waterloo and victor of Austerlitz, or equiangular and

---

* Such an individual name is no longer a concept word: it can no longer be used with the indefinite article or without article.

---

4 Linke notes:
  Frege's 'concept' (predicative function) seems to me to overlook that the predicate also contains the problem of generality. There is something non-individual (= not here and now) which has nothing whatsoever to do with, e.g., a 'class'. In the above example, it would *not* hold for it that it was not blue, etc. L.

equilateral triangle.' Here you throw together different things.[5] 'Loser of Waterloo' is a *nomen appellativum*, and so is 'victor of Austerlitz'. They both designate concepts and not objects. But we can form individual names out of them: 'the loser of Waterloo' and 'the victor of Austerlitz'. 'Equiangular triangle' does not designate an object, nor can we form an individual name out of it by means of the definite article. This is why I would not use the expression 'the object meant by the concept equiangular triangle'. Here it is not an object which is identical with an object, but a concept which is coextensive with a concept. We cannot properly speak of identity in the case of concepts.

You write, 'The characteristic mark product does not fit 7 plus 5 at all.' 7 plus 5 presents itself as a proper name. In case it achieves its purpose, it means an object. According to my way of using the words 'characteristic mark' and 'property', there can be no talk here of a characteristic mark, only of properties.

Now the property of being a product cannot indeed be known immediately from the designation '7 plus 5'; but the property of being victor of Austerlitz cannot be gathered either without further ado from the designation 'the loser of Waterloo'. We can indeed know this or that property of the designated object from the individual name; but there is no reason why all the properties of the designatum should be capable of being known in this way. Why could the loser of Waterloo not have properties which are not characteristic marks of the concept loser of Waterloo? Let us compare the proposition '$7 + 5 = 12$', or in words, 'the sum of 7 and 5 is 12',[6] with the proposition '$7 + 5$ is an even number'. From a purely linguistic point of view the two cases seem to be the same; but as in many other cases language disguises the state of affairs. In the second proposition, the 'is' is used in the ordinary way as a copula. It can be very clearly recognized as such when the linguistic predicate consists of 'is' and an adjective or of 'is' and a concept word either without article or with the indefinite article ('is an even number'), while the grammatical subject is the name of an object ('$7 + 5$' or 'the sum of 7 and 5'). We then have subsumption of an object under a

---

5 Linke notes:
Today I should say: concepts and hence, at the same time, (*delimited*) objects; concepts only, if like Frege one belongs to the nominalist tradition. Otherwise 'victor of Waterloo' [perhaps 'Austerlitz' is meant] designates an object, or more precisely, a non-individual character (today I should no longer speak of characteristic marks), similar to the spherical character of this 'sphere' for example, if we regard everything that is individual and unique about it as irrelevant (or better: to the extent that its individuality and uniqueness is inessential within a certain 'objective' whole). What is spherical about this (perhaps previously many-cornered thing) corresponds exactly to 'loser of Waterloo'. L.

6 Linke notes:
But the predicate does not read '12' but 'equal to 12'.

concept. This is quite different from the case where the grammatical predicate consists of 'is' and the name of an object ('12'). The two object names between which 'is' stands are then interchangeable; thus '7 + 5 = 12' can be turned into '12 = 7 + 5'. We then have an equation. In this case 'is' has the sense of the equals sign and contains an essential part of the sense of the predicate, while as a copula it has no sense of its own and only serves to mark the predicate as such. In the proposition 'Napoleon is loser of Waterloo' we have the sense of subsumption;[7] in the proposition 'Napoleon is the loser of Waterloo' we have an equation. The left designatum is the same as the right designatum, just as in the proposition 'the loser of Waterloo is the victor of Austerlitz', or as in the proposition '7 + 5 is 6 · 2'. But even though these object names have the same meaning, they have a different sense, and this is why the proposition '7 + 5 is 7 + 5', '7 + 5 is 12', and '7 + 5 is 6 · 2' express different thoughts; for the sense of part of a proposition is part of the sense of the proposition, i.e., of the thought expressed in the proposition.

If a thought is an equation, this does not exclude its also being a subsumption. The proposition 'Napoleon is the loser of Waterloo' can be changed into 'Napoleon is identical with the loser of Waterloo'; and here Napoleon is subsumed, not indeed under the loser of Waterloo, but under the concept 'identical with the loser of Waterloo'.

You now continue: 'Only if I consider 7 + 5 and 2 · 6 with regard to a very definite characteristic mark, namely the numerical value they both have, only then can I speak of the identity of the two.'

Now on my view the numerical value is not a characteristic mark of 7 + 5 or a property of it, but this numerical value is 7 + 5 itself.[8] There is only one sum 7 + 5, and I designate it by 'the sum of 7 and 5' or by '7 + 5' or also by '6 · 2'. I can of course say that '7 + 5' – that is, this combination of signs – has a numerical value, but only as its meaning, not as a characteristic mark.

I hope I have not bored you with these explanations, which turned out to be somewhat didactic. I needed to discuss this question thoroughly for once.

With best regards,

Yours sincerely,
G. Frege

---

7 Linke notes:
  What is *designated* by the word 'Napoleon' and what has the character 'loser of Waterloo' is the same; or also: the wholes formed in the one case by the name and the object designated by it and, in the other, by the character and the object that possesses it have something in common, namely the object. L.

8 Linke notes:
  If this was so, I would not know *more*, or learn *more*, when I learn that the sum of 7 and 5 is equal to the number 12 (i.e., that this sum and the 12th number in the number series have the number in common ⟨= the same number⟩). L.

XII FREGE–MARTY

Editor's Introduction

Anton Marty (1847–1914) was a pupil of Brentano and, from 1890 on, professor of philosophy at Prague University. Apart from the following letter from Frege, no correspondence between Frege and Marty has been found, and there is some question whether the following letter was really addressed to Marty. The attribution to Marty is based on a letter from Prof. G. Katkov (then of Prague) to H. Scholz dated 4.2.1938, which gives Marty as the addressee. But the original of the letter, which was then in the Brentano Archives in Prague, is now lost, and if one compares its contents with the contents of Stumpf's letter to Frege (XVI/1), it appears that Stumpf's letter is on all points a reply to Frege's letter to Marty. Since Marty and Stumpf were colleagues at Prague in 1882 and, as pupils of Brentano, close collaborators, one cannot exclude the possibility that Katkov mistook a letter to Stumpf for a letter to Marty. On the other hand, the attribution to Marty is supported by the fact that Marty seems to have acceded to Frege's request to 'call attention' to the *Begriffsschrift* in a journal. In the second of a series of seven articles Marty goes into Frege's theory of judgement.[1] Marty agrees with Frege's view that from a logical point of view there is no difference between subject and predicate. On the other hand he rejects Frege's treatment of negation as part of the 'judgeable content'. Marty takes the view that negation and affirmation are 'coequal forms of the function of judgement'.[2] In turn, Frege takes a position on Marty's views, though only indirectly, by commenting to Husserl on Husserl's critique of Marty.[3]

XII/1 [xxx/1]   FREGE to MARTY   29.8.1882

Jena
29 August 1882

Dear Colleague,

Your friendly postcard gave me much pleasure, the more so as I have found only very little agreement up to now. Allow me to give you some more information about my *Conceptual Notation*, in the hope that you will perhaps have occasion to call attention to it in a journal; it would make it easier for me to publish further works. I have now nearly completed a book in which I treat the concept of number and demonstrate that the first principles of computation which up to now have generally been regarded as unprovable axioms can be proved from definitions by means of logical laws alone, so that they may have to be regarded as analytic judgements in

---

1 The second article is 'Über subjektlose Sätze und das Verhältnis der Grammatik zur Logik und Psychologie', *Vierteljahrsschrift für wissenschaftliche Philosophie* 8 (1884) pp. 161–92. The discussion of Frege is on pp. 185–8.
2 ibid., p. 188.
3 Cf. VII/3 above as well as notes 2 and 3 to VII/3.

Kant's sense.[1] It will not surprise me and I even expect that you will raise some doubts about this and imagine that there is a mistake in the definitions, in that, to be possible, they presuppose judgements which I have failed to notice, or in that some other essential content from another source of knowledge has crept in unawares. My confidence that this has not happened is based on the application of my conceptual notation, which will not let through anything that was not expressly presupposed, even if it seems so obvious that in ordinary thought we do not even notice that we are relying on it for support. This seems to place the value and the power of discursive thought in the right light. For whereas Leibniz may well have overestimated it when he hoped to prove everything from concepts, Kant on the contrary seems to me to place too low an estimate on the significance of analytic judgements because he sticks to over-simple examples. I regard it as one of Kant's great merits to have recognized the propositions of geometry as synthetic judgements, but I cannot allow him the same in the case of arithmetic. The two cases are anyway quite different. The field of geometry is the field of possible spatial intuition; arithmetic recognizes no such limitation. Everything is enumerable, not just what is juxtaposed in space, not just what is successive in time, not just external phenomena, but also inner mental processes and events and even concepts, which stand neither in temporal nor in spatial but only in logical relations to one another. The only barrier to enumerability is to be found in the imperfection of concepts. Bald people for example cannot be enumerated as long as the concept of baldness is not defined so precisely that for any individual there can be no doubt whether he falls under it. Thus the area of the enumerable is as wide as that of conceptual thought, and a source of knowledge more restricted in scope, like spatial intuition or sense perception, would not suffice to guarantee the general validity of arithmetical propositions. And to enable one to rely on intuition for support, it does not help at all to let something spatial represent something non-spatial in enumeration; for one would have to justify the admissibility of such a representation. But I wanted to tell you something about my conceptual notation. You emphasize the division between the function of judgement and the matter judged. The distinction between individual and concept seems to me even more important. In language the two merge into each other. The proper name 'sun' becomes a concept name when one speaks of suns, and a concept name with a demonstrative serves to designate an individual. In logic, too, this distinction has not always been observed (for Boole only concepts really exist). The relation of sub-ordination of a concept under a concept is quite different from that of an individual falling under a concept. It seems that logicians have clung too

---

XII/l. 1 The book referred to is GLA. It seems that Frege followed C. Stumpf's suggestion (cf. XVI/1) and decided to give an analysis of the concept of number without using his conceptual notation. This is not what he had originally planned to do, as shown by this letter.

much to the linguistic schema of subject and predicate, which surely contains what are logically quite different relations. I regard it as essential for a concept that the question whether something falls under it have a sense. Thus I would call 'Christianity' a concept only in the sense in which it is used in the proposition 'this (i.e., this way of acting) is Christianity', but not in the proposition 'Christianity continues to spread'. A concept is unsaturated in that it requires something to fall under it; hence it cannot exist on its own. That an individual falls under it is a judgeable content, and here the concept appears as a predicate and is always predicative. In this case, where the subject is an individual, the relation of subject to predicate is not a third thing added to the two, but it belongs to the content of the predicate, which is what makes the predicate unsatisfied. Now I do not believe that concept formation can precede judgement because this would presuppose the independent existence of concepts, but I think of a concept as having arisen by decomposition from a judgeable content. I do not believe that for any judgeable content there is only one way in which it can be decomposed, or that one of these possible ways can always claim objective pre-eminence. In the inequality $3 > 2$ we can regard either 2 or 3 as the subject. In the former case we have the concept 'smaller than 3', in the latter, 'greater than 2'. We can also regard '3 and 2' as a complex subject. As a predicate we then have the concept of the relation of the greater to the smaller. In general I represent the falling of an individual under a concept by $F(x)$, where x is the subject (argument) and $F(\ )$ the predicate (function), and where the empty place in the parentheses after F indicates non-saturation. The subordination of a concept $\varPsi(\ )$ under a concept $\varPhi(\ )$ is expressed by

$$\begin{array}{c} \multimap^{\mathfrak{a}} \!\!\!\begin{array}{l} \varPhi(\mathfrak{a}) \\ \varPsi(\mathfrak{a}) \end{array} \end{array}$$

which makes obvious the difference between subordination and an individual's falling under a concept. Without the strict distinction between individual and concept it is impossible to express particular and existential judgements accurately and in such a way as to make their close relationship obvious. For every particular judgement is an existential judgement.

$$\vdash\!\!\smile^{\mathfrak{a}}\!\!\top \mathfrak{a}^2 = 4$$

means: 'There is at least one square root of 4'.

$$\vdash\!\!\smile^{\mathfrak{a}}\!\!\top\begin{array}{l} \mathfrak{a}^2 = 4 \\ \mathfrak{a}^3 = 8 \end{array}$$

means: 'Some (at least one) cube roots of 8 are square roots of 4'. One can insert here two negation strokes that cancel each other:

$$\vdash\!\!\smile^{\mathfrak{a}}\!\!\top\!\!\top\begin{array}{l} \mathfrak{a}^2 = 4 \\ \mathfrak{a}^3 = 8 \end{array}$$

and think of this as composed of 1, $\tau\overset{\mathfrak{a}}{\smile}\tau$, and

$$\text{⊤}\begin{array}{l} \mathfrak{a}^2 = 4 \\ \mathfrak{a}^3 = 8 \end{array}.$$

The latter takes the place of $\mathfrak{a}^2 = 4$ in

$$\vdash\overset{\mathfrak{a}}{\smile}_\tau \mathfrak{a}^2 = 4$$

so that it can be translated as: 'There is at least one number which is both a cube root of 8 and a square root of 4'. Existential judgements thus take their place among other judgements. I should still like to show you how Kant's refutation of the ontological argument becomes intuitively very obvious when presented in my way[2] and what the value of the concavity is, which is my sign of generality, but I fear I have already overburdened you with my long letter. I find it difficult to gain entry into the philosophical journals. Please excuse this letter as springing from my unsatisfied need for communication. I find myself in a vicious circle: before people pay attention to my conceptual notation, they want to see what it can do, and I in turn cannot show this without presupposing familiarity with it. So it seems that I can hardly count on any readers for the book I mentioned at the beginning. If you would be so good as to answer me, I would ask you to communicate your doubts. I should like to find out what you think of the scientific value of the demonstration I am planning, supposing it succeeds and is carried out with the most painstaking precision.

Yours sincerely,

G. Frege

---

2 Frege's conceptual notation makes it clear that existence is only a property of a (first-level) concept and hence itself a second-level concept, to use Frege's terms. Existence cannot therefore be a characteristic mark of a first-level concept, e.g., of the concept 'God', nor, therefore, can it be discovered by analysis within it.

XIII FREGE–PASCH

Editor's Introduction

Moritz Pasch (1843–1930) was professor of mathematics at Giessen at the time of his correspondence with Frege. Pasch especially distinguished himself in his investigations into the axiomatization of geometry. If Frege sent Pasch some of his writings, especially on the foundations of arithmetic and geometry, it was probably because Pasch was one of the few colleagues of Frege to take a serious interest in the logical and methodological problems of mathematics.[1]

XIII/1 [xxxiii/1]   PASCH to FREGE   11.2.1894

Giessen
11 February 1894

Dear Colleague,

Since I have been much concerned with the foundations of arithmetic, I was anxious to get to know your book, *The Basic Laws of Arithmetic* I, 1893. I have now started to read it; but to my regret I must – by reason of official business – forego further study of it for some time. I was interested to see that you discuss in depth the nature of mathematical proof, a subject on which there are indeed few clear ideas around. In particular, I must agree whole-heartedly with your demand that presuppositions to be added tacitly in thought ought not to be tolerated. This is what I tried to do in geometry in 1882, in my book *Lectures on the Newer Geometry*,[1] although I have had some second thoughts about its starting-points. In case you should be interested, allow me to refer also to the short essays in *Mathematische Annalen*, vols 30, 32 and 40.[2] Arithmetic must be placed on firm foundations before this can be done for geometry. In this respect I have not yet been able to make up my mind to regard arithmetic as merely part of logic.

But I must not go deeper into this subject now, and I conclude by asking you for a friendly reception of these lines.

Yours sincerely,
M. Pasch

---

1 On Pasch's life and work, cf. his autobiography: Moritz Pasch, *Eine Selbstschilderung* (Giessen 1930), as well as F. Engel and M. Dehn, 'Moritz Pasch', *Jahrbuch der deutschen Mathematiker-Vereinigung* 44 (1934), pp. 120–42.
XIII/1. 1 *Vorlesungen über neuere Geometrie* (1882).
2 'Ueber die projective Geometrie und die analytische Darstellung der geometrischen Gebilde', *Mathematische Annalen* 30 (1887), pp. 127–31; 'Ueber einige Punkte der Functionentheorie', ibid., pp. 132–54; 'Ueber die uneigentlichen Geraden und Ebenen', *Mathematische Annalen* 32 (1888), pp. 159ff; 'Ueber die Einführung der irrationalen Zahlen', *Mathematische Annalon* 40 (1892), pp. 149–52.

XIII/2 [xxxiii/3]   PASCH to FREGE   14.10.1896

Giessen
14 October 1896

Dear Colleague,

I have received the valuable package you mention on your postcard of 13 October, and I thank you very much for it.[1] If there exist differences between you and Peano, it should afford you some satisfaction to know that your writings and teachings are being taken more and more into account.[2]

Yours sincerely,
M. Pasch

XIII/3 [xxxiii/4]   PASCH to FREGE   14.11.1899

Giessen
14 November 1899

Dear Colleague,

You were so kind as to send me your paper on Schubert's numbers.[1] I have read it with great interest, the more so as I completely share your negative feelings about certain phenomena in the mathematical literature. But how difficult it is to remedy this state of affairs! If our mathematical books contain no satisfactory answer to any, or hardly any, of the fundamental questions, it is to my mind only partly due to the incompetence or indifference of their authors. Too much material has been accumulated, and if one tries to sift it, one gets bogged down at every step. An individual can do only little about this. When you exclaim ironically: How many forces are deflected by such questions from the important things!,[2] there is some truth in this thought. If considerations like the ones you pursue – which I have no hesitation in regarding as urgently desirable – lead one to sharpen one's vision for gaps in definitions and proofs, one will again and again be forced to think about the deeper questions and must take time off from the proper subject matter. But what will then become of the activity of teaching? The books that advance without a single doubt prove to be most successful with students – much error and a tiny spark of truth. This recognition may be painful, but one cannot shut one's eyes to the fact.

_____

XIII/2. 1 The parcel may have contained offprints of Frege's letter to Peano ⟨14⟩, published in 1896 (cf. XIV/7 below), and of Frege's lecture on Peano's notation and his own ⟨15⟩, published in 1897.
2 On the differences between Frege and Peano, which are mainly connected with the use of Peano's implication sign, '⊃', cf. Frege's 'The Argument for my stricter Canons of Definition', PW, pp. 152–6, as well as ⟨14⟩ and ⟨15⟩ and the Frege–Peano correspondence (XIV below).
XIII/3. 1 = ⟨16⟩.
2 ibid., p. IV.

Many take the position that mathematics only has to draw conclusions from a certain material (say the content of elementary mathematics) and that it is not up to the mathematician to criticize it. This is at least an understandable position. It is perhaps from this position that many reputable mathematicians bring themselves to acknowledge works from which one feels like turning away in horror. I wish you the greatest success in your enterprise, the forceful advocacy of genuine science. It seems to me that by the clarity and intelligibility of your presentation you have happily overcome the difficulty that the subject presents. In asking you to accept my most sincere thanks for the valuable paper you sent me, I remain,

Yours sincerely,
M. Pasch

XIII/4 [xxxiii/5]   PASCH to FREGE   18.1.1903

Giessen
18 January 1903

Dear Colleague,

The book you kindly sent me, which it was an honour to receive,[1] reached me at a time when I was deeply concerned with the object of your investigations because I had to discuss it in a lecture. For this reason alone, I immediately picked up your book and read it, as much as this is possible without a mastery of your conceptual notation; the latter is impossible for me, given my age and the heavy demands on my time. While there is a widespread feeling that the foundations of analysis are in need of clarification, the execution of this task is in poor shape. And yet one cannot undertake the corresponding task for geometry and mechanics till the other is completed. You, dear colleague, are one of the few to have pursued this matter in depth and perhaps the only one to have stayed strictly aloof from conventional verbiage. Your criticism of existing efforts seems to me completely justified, although in part I cannot agree with the reasons you advance against them. In general, I believe I cannot accept your positive claims without qualification. It is unfortunately impossible for me to explain this more fully within the framework of a letter and given the severe limits on my time. But I should at least like to raise the question, against the requirements you lay down for definitions, pp. 67ff [E.T. pp. 159ff], how you propose to explain, e.g., infinitely distant points.

Please receive my most sincere thanks for kindly sending me your very valuable book, and be so kind as to forgive me if my letter does not go in greater detail into the wealth of material it contains. I should be very glad if I could make up for this on a later occasion.

With the wish that your work may find many serious readers, I remain,

Yours sincerely,
M. Pasch

XIII/4. 1 GGA II.

XIII/5 [xxxiii/7]   PASCH to FREGE   7.1.1905

Giessen
7 January 1905

Dear Colleague,

You were so kind as to return to a remark I made on 18 January 1903 and to discuss in depth the introduction of infinitely distant elements even though I had not yet replied to your letter of 18 April 1904.[1] What happened to me was like what you wrote about yourself. I have been absorbed in other things, and it is so hard to tear oneself away. Besides, I cannot write you anything satisfactory. Your procedure is in itself logically correct and complete; but I cannot reconcile it with my position. It seems to me that the concept 'class', e.g., class of Euclidean lines parallel to a Euclidean line, is already in need of definition, and I do not know how it is to be defined if one is not allowed to exercise the right to define a whole phrase. In my *Geometry* of 1882 I defined the ideal elements in this sense – and for the case of non-Euclidean geometry as well.[2] Although I do not now regard the presentation I then adopted as precise enough in its details, I should like to maintain it on the whole. In the case of non-Euclidean geometry, your definitions would have to undergo considerable alterations. In the case of Euclidean geometry, the infinitely distant point *c* and the class *c* of Euclidean lines parallel to a certain Euclidean line becomes for you one and the same thing; in other words, the class of Euclidean lines going through *c* is *c* itself. This may well give rise to intolerable consequences.

You will not allow one to talk first about 'points' because a 'point' is later to be understood as something broader, not just as a Euclidean point. But even after the introduction of Euclidean and infinitely distant points the matter is not concluded: imaginary points are still to come. I think that your intention, which I find perfectly evident, will also be fulfilled if one talks first about 'points' without qualification and then extends the application of the

---

XIII/5. 1 According to a fragmentary draft of this letter, which is now lost, Frege proposed to define infinitely distant points by *abstraction*, that is, in each case as classes of all lines parallel to a certain line. This introduction of the term would correspond in effect to the definition of the direction of a line in GLA, sect. 68: 'The direction of line *a* is the extension of the concept "parallel to line *a*"'. If we conceive extensions of concepts as classes, then 'the infinitely distant point on line *a*' is only another way of speaking of 'the direction of *a*'.

2 In *Vorlesungen über neuere Geometrie*, Pasch arrives at infinitely distant points in a way similar to the way proposed by Frege: he correlates statements about a point with statements about a 'cone of rays' consisting of lines going through this point, and he continues this correlation by also correlating talk about infinitely distant points with talk about 'improper' cones of rays consisting of lines parallel to one another. According to what he says above in the letter, he avoids conceiving of cones of rays as abstract objects, and he confines himself by and large to the phrase '. . . is a ray of the cone of rays determined by two lines *e* and *f*'.

word 'point' while stating that in all preceding propositions and definitions the word 'point' is to be understood only in the original sense. This is what I did in the work I cited, and this is what others have done, even if not always with the necessary sharpness. You have stated repeatedly how much mathematical literature is lacking in sharpness. There is a strong disinclination towards true clarity, which is really incomprehensible, but is evidently founded in human nature. Your explanations were all the more valuable to me, and I thank you very much for them.

Yours sincerely,
M. Pasch

XIII/6 [xxxiii/8]   PASCH to FREGE   13.11.1906

Giessen
13 November 1906
Dear Colleague,

I have been following with interest the expressions of opinion on the fundamental questions of mathematics, which have led to lively exchanges in very recent times. Your interventions, to which the paper you kindly sent me bears witness, serve to advance those questions in a gratifying way.[1] They sharpen the mathematician's conscience, which tends to be somewhat slack in this respect. While I believe that I cannot adopt your position on everything, I must for lack of time forego a closer investigation and exposition of this. Instead I must confine myself to the expression of my warmest thanks, with which I am,

Yours sincerely,
M. Pasch

XIII/6. 1 The paper in question may have been GLC III 1–3 or ⟨24⟩.

Editor's Introduction

Giuseppe Peano (1858–1932) was professor of mathematics at Turin. The best known of his many contributions to logic is the *Formulaire de mathématiques* (1895–1905), in which he developed a symbolism which, through the mediation of Whitehead and Russell's *Principia Mathematica*, became the basis of the logical symbolisms employed today. Except for a postcard from the year 1903, the surviving correspondence dates from the years 1894–6, when Frege had already published GGA I and while Peano was preparing *Formulaire de mathématiques* II for publication.

Peano wrote to Frege in French, but the letter here numbered XIV/8 is preserved in and translated from an Italian publication. Frege's letters were in German.

XIV/1 [xxxiv/1]   FREGE to PEANO   undated[1]

Dear Colleague,

Your papers and your postcard gave me great pleasure, and I thank you very much for them. Unfortunately I have no knowledge of the Italian language. Still, I have tried to understand your essays, and I believe I have succeeded in grasping the essentials of your notation. I hope to read myself more and more into Italian. I am therefore looking forward with pleasure to the promised parcel as well, and I thank you very much for it in advance.

I see that you follow Boole, but use some other signs and try to make his logic fruitful for mathematics. You too have found it necessary to introduce a designation for generality ($\supset_{x,\,y}$, $=_{x,\,y}$) and use it to reproduce our 'there is' (existence).[2] Compare this with my '$\underset{\mathfrak{a}}{\mathsf{h}\smile\mathsf{T}}\, \mathfrak{a}^2 = 1$' (There are square roots of one). This agreement in substance combined with a total difference in form is especially valuable to me since with most logicians there seems to prevail great unclarity about the nature of these existential judgements, and this is true of the followers of Boole as well as of the psychological logicians, even though Kant seems already to have been on the right track in his criticism of the ontological argument for the existence of God. Your designation for generality remedies an essential defect of Boolean logic; but it is perhaps less generally applicable than mine, and I do not know whether you will be able to set sure limits to the scope of generality in all cases. You use the letters x and y to restrict the generality to part of a proposition, i.e., where I employ German letters. But if the scope of generality is to comprise the whole

---

XIV/1. 1 This letter cannot have been written before 1891, since it refers to one of Peano's articles published that year. Frege's remark that he is ignorant of Italian suggests that this is his first letter to Peano.

2 In this, as in the German edition of the correspondence, the sign '$\supset$' has been substituted for the sign 'O' which Frege and Peano used in the original and which is a reversed capital 'C' for *contient* ('contains'). This substitution is in accord with Peano's later introduction of '$\supset$' for 'O'.

proposition, you make use of the other Roman letters, just as I do. Thus proposition P6 in sect. 2 of your 'On the Concept of Number' is to be valid in general whatever a may be.[3] Of course, where you explain your signs but do not yet apply them, you use the Roman letters also in another way, i.e., as I use Greek capital letters, but this is an incidental matter.

Closely connected with generality is my extended application of functional indications by means of the letters f, F, etc., and I believe that you will come around to making use of this unless you are already doing so. Then your '$f(x) \supset_x F(x)$' would correspond to my

$$\overset{a}{\cup}\mathbf{\top}\begin{matrix} F(\mathfrak{a}) \\ f(\mathfrak{a}) \end{matrix}$$

It seems to me very important that you make a sharp distinction between singular and universal propositions by means of the signs ε and $\supset$, since the relations of an object (individual) to a concept under which it falls and to the class to which it belongs are indeed quite different from the relation of a concept to a superordinate concept or of a class to a more comprehensive class. This difference is overlooked by many writers, as by Mr Dedekind in his work *The Nature and Purpose of Numbers*.[4] Its importance comes out especially when empty concepts and classes are taken into consideration. For these must be recognized, and you actually do this when you use the sign Λ. Of course, one must not then regard a class as constituted by the objects (individuals, entities) that belong to it; for in removing the objects one would then also be removing the class constituted by them. Instead one must regard the class as constituted by the characteristic marks, i.e., the properties which an object must have if it is to belong to it. It can then happen that these properties contradict one another, or that there occurs no object that combines them in itself. The class is then empty, but without being logically objectionable for that reason. Now from the singular proposition 'One is a fourth root of one' ($\vdash 1^4 = 1$) one can indeed infer: 'There are fourth roots of one' ($\vdash\overset{a}{\cup}\mathbf{\top} \mathfrak{a}^4 = 1$), but from the universal proposition 'All square roots of one are fourth roots of one' . . .[5]

XIV/2 [xxxiv/2]   PEANO to FREGE   30.1.1894

Turin
30 January 1894

Dear Colleague,

Thank you for the notes you sent me. Some time ago I bought your *Foundations of Arithmetic*,[1] and I got our library to buy your recent *Basic*

---

3 'Sul concetto di numero', *Rivista di matematica* 1 (1891), pp. 87–102 and 256–67.
4 R. Dedekind, *Was sind und was sollen die Zahlen* (Brunswick 1888) [E.T. in *Essays on the Theory of Number* (Chicago 1901; New York 1963)].
5 The draft of the letter ends at this point.
XIV/2. 1 = ⟨4⟩.

Laws of Arithmetic.[2] I have some difficulty in reading your symbols; but I shall get better at it, and if I still find difficulties, I shall take the liberty of writing to you. From the notes I sent you, and from one that is being printed and which you will receive in a few days, you will see that we are taking the same route in science. If you find some difficulty in my notes, it will give me pleasure if you communicate it to me.

Yours sincerely,
G. Peano
Professor at the University

XIV/3 [xxxiv/3]   PEANO to FREGE   10.2.1894

Turin
10 February 1894

Dear Colleague,

I am sending you my latest publication, *Notations of Mathematical Logic*,[1] and I am adding the *Principles of Arithmetic*, of which I was still able to find copies.[2]

In my latest publication, which serves as an introduction to the *Formulary of Mathematics*,[3] I propose to give a clear and simple explanation of the notations of logic or, if you prefer, conceptual notation. I hope it will serve to make known these new theories, which, unfortunately, are still little known and even despised. You who have contributed so much to the formation and development of these theories can also contribute to their diffusion by publishing a critique, or review, of my work.

In my article you will find many ideas which you have published in your works. But on several points our views are different.

I am reading your articles with ever-increasing pleasure, but slowly, for I am very busy with my teaching and with some other projects, and I am still not familiar with all of the notation you have adopted in your latest book. Nevertheless I shall ask you for an explanation:

How do you write the following propositions?

(1) There is a whole positive number x which satisfies the equation
$$2 + 3x = 5 \qquad (x \, \varepsilon \, \mathrm{N} \cdot 2 + 3x = 5 \cdot \sim =_x \Lambda)$$

(2) There is no whole positive number x which satisfies the condition
$$2 + 3x = 6 \qquad (x \, \varepsilon \, \mathrm{N} \cdot 2 + 3x = 6 \cdot =_x \Lambda)$$

$2 = \langle 11 \rangle$.

XIV/3. 1 G. Peano, *Notations de logique mathématique: Introduction au Formulaire de Mathématique* (Turin 1894).

2 G. Peano, *Arithmetices principia, nova methodo exposita* (Turin 1889).

3 G. Peano, *Formulaire de mathématiques* I (Turin 1895).

(3) There is a rational number which satisfies the above condition
$$(x \, \varepsilon \, R \cdot 2 + 3x = 6 \cdot \sim=_x \Lambda)$$
etc.

The propositions in parentheses are written in the symbols I am proposing. In your book I did not find a sign to indicate that x is whole, or rational, or real, or imaginary, etc.[4]

Yours sincerely,

G. Peano

Professor at the University of Turin

XIV/4 [xxxiv/4]   PEANO to FREGE   14.8.1895

Pilonetto (Turin)

14 August 1895

Sir,

Some months ago I sent you the first volume of the *Formulary*[1] and subsequently some of my articles on mathematical logic.

I have finished reading your book, and I am preparing a review of it.[2] Is the second volume of your work in the press?

Yours sincerely,

G. Peano

XIV/5 [xxxiv/5]   PEANO to FREGE   24.10.1895

Thank you for your article from the *Archiv*.[1] Your reasons seem to me well taken, and I shall read your whole work attentively. The question

---

4 Frege wrote down the following answers to Peano's questions on the last page of Peano's letter:

XIV/4. 1 Cf. n. 3 to XIV/3.
2 The review appeared in *Rivista di matematica* 5 (1895), pp. 122–8.
XIV/5. 1 = ⟨13⟩.

concerning the distinction between the signs $\supset$ and ε also bears on the introduction of the sign ι ($\iota\sigma o\varsigma$: my *Introduction*, sect. 31), for we have (*Formulary* I, sect. 4, P14):

$$a\,\varepsilon\,\mathrm{K}\cdot\supset:x\,\varepsilon\,a\cdot=\cdot\iota\,x\supset a$$

You also sent me an issue of the *Revue de métaphysique*,[2] but I did not understand your reason for sending it. Did you receive the fascicles of the *Rivista* containing the review of your book?

Some months ago you promised me a review or some other kind of work on the *Formulary* we are publishing.[3] Up to now I have received nothing.

Your collaboration on the *Rivista di matematica* would be very valuable to us, no matter what its form.

Yours sincerely,
G. Peano

XIV/6 [xxxiv/6]   PEANO to FREGE   5.4.1896

Pilonetto (Turin)
5 April 1896

Dear Colleague,

As you wrote to me, and as I have seen announced, you talked about our symbolisms before the German Association of Mathematical Sciences.[1] But till now I have not yet been able to read your article. Would you send me a copy of it? The questions we are concerned with are on the one hand very interesting and on the other quite difficult. We all share the desire to clarify them, to discuss them, and to further their solutions.

Yours sincerely,
G. Peano
Professor at the University of Turin

XIV/7 [xxxiv/7]   FREGE to PEANO   29.9.1896

Jena
29 September 1896

Dear Colleague,

You were so good as to write a detailed and favourable review of my *Basic Laws of Arithmetic* in vol. 5 of the *Rivista di matematica*.[1] I believe

---

2 = $\langle 12 \rangle$.

3 Cf. $\langle 15 \rangle$.

XIV/6. 1 The reference is presumably to a lecture Frege gave in the Mathematical Division of the Scientific Congress in Lübeck, an expanded version of which was presented to the Special Session of the Royal Saxon Society for the Sciences in Leipzig on 6.7.1896. Cf. $\langle 15 \rangle$.

XIV/7. 1 *Rivista di matematica* 5 (1895), pp. 122–8.

that I can best show you my gratitude if I candidly explain where my views diverge from yours. Such an exchange of ideas will, I hope, advance our science, especially if carried on in public; and I should therefore like to ask you to publish the following explanations in the *Rivista di matematica*.[2]

You say on p. 123:

'It is well known that in the *Formulary of Mathematics* all the relations and operations between propositions and between classes are reduced to three fundamental ones which are indicated by the signs

$$\cap, \supset, -.$$

Besides these, there are used, for convenience, the signs =, $\cup$, and Λ, which are defined by means of the preceding ones.'

And on p. 125:

'The two systems of notation can be compared from the scientific and the practical points of view. From the scientific point of view, Frege's system is based on five fundamental signs:

$$\mathsf{I}, \overline{}, \overline{\top}, \underset{\llcorner}{\top}, \overline{\smile}$$

while ours is based on three signs:

$$-, \cap, \supset.$$

Hence the system of the *Formulary* represents a more profound analysis.'

The last point I should not like to admit. To begin with, I doubt whether you can really designate all the purely logical forms you use by those three signs. I have my doubts even about identity. In *Formulary* I, sect. 1, 3, we find a definition of the sign = in the form

$$a = b \cdot = \cdot a \supset b \cdot b \supset a.$$

But this evidently presupposes the meaning of the definiendum. For in the Preface, p. iv, you say: 'Every definition is expressed by an equation; the first term is the sign that is being defined.' Here the left side of the definitional equation is '$a = b$', the right side '$a \supset b \cdot b \supset a$', and between the two occurs the sign of identity whose meaning must therefore be already known if the definition is to be understood. For this reason alone there can be no question of a reduction of identity to the meanings of the three primitive signs. But the definition is objectionable also in other respects. It is not the only definition by which the sign of identity is defined: in *Formulary* I, sect. 4, 2, there follows a second, in sect. 5, 2, a third, and finally in sect. 5, 11, a fourth definition. This reveals a fundamental opposition between our views. I reject multiple definitions of the same sign for the following reason.

---

2 This letter was published in *Rivista di matematica* 6 (1896–9), pp. 58–9, together with Peano's reply (XIV/7 below), ibid., pp. 60ff.

Let us suppose that we have two definitions and that both of them give a meaning to the same sign. Then there are only two conceivable cases: either both of them give the same meaning to the sign, or not. In the first case, we have again two possibilities: either both definitions bestow the same sense on the sign and say exactly the same, or not. In the first case, one of the two is superfluous; in the other, it would have to be proved that they assign the same meaning to the sign even though they give it different senses. Thus one of the two would have to be allowed to stand as a definition while the other would have to be changed into a theorem and proved. The reader is cheated of this proof if what should be a theorem is presented as a definition. Finally, if the definitions give different meanings to the same sign and not just different senses, then they contradict each other, and one of the two must give way. It must strike us now that none of these possibilities seems to be realized in our case; for your definitions of the sign of identity do not all say the same, nor can one be proved from the other, nor do they contradict one another. The explanation for this is that one of the presuppositions we just made is not satisfied. Namely, it is not true to suppose that each of these definitions gives a meaning to the sign of identity. For every one of them is incomplete, and only a complete definition can assign a meaning to a sign. Your first definition (I, sect. 1, 3) evidently refers to the case where the sign of identity connects propositions, even though this is not stated explicitly; the second (I, sect. 4, 2) envisages the case where it occurs between signs for classes; the third (I, sect. 5, 2) the case where it equates individual things, and the fourth (I, sect. 5, 11) the case where it connects function signs. What case is meant is stated as a rule by a conditional proposition. Now I reject conditional definitions, which you frequently give, because they are incomplete, because they only state for certain cases, not for all, that the new expression is to mean the same as the explaining one. And so they miss their aim of given a meaning to a sign. Why is this so? Let us consider the cases where we are dealing with a concept and where we are dealing with a relation as in the case of identity. A conditional definition of the sign for a concept decides only for some cases, not for all, whether an object falls under the concept or not; it does not therefore delimit the concept completely and sharply. But logic can only recognize sharply delimited concepts. Only under this presupposition can it set up precise laws. The logical law that there is no third case besides

$$a \text{ is } b$$

and

$$a \text{ is not } b$$

is really only another way of expressing our requirement that a concept (*b*) must be sharply delimited. The fallacy known by the name of 'Acervus' rests on this, that words like 'heap' are treated as if they designated a sharply delimited concept whereas this is not the case. Just as it would be impossible for geometry to set up precise laws if it tried to recognize threads as lines and knots in threads as points, so logic must demand sharp limits of what it will recognize as a concept unless it wants to renounce all precision and certainty. Thus a sign for a concept whose content does not satisfy this requirement is to be regarded as meaningless from the logical point of view. It can be objected that such words are used thousands of times in the language of life. Yes; but our vernacular languages are also not made for conducting proofs. And it is precisely the defects that spring from this that have been my main reason for setting up a conceptual notation. The task of our vernacular languages is essentially fulfilled if people engaged in communication with one another connect the same thought, or approximately the same thought, with the same proposition. For this it is not at all necessary that the individual words should have a sense and meaning of their own, provided only that the whole proposition has a sense. Where inferences are to be drawn the case is different: for this it is essential that the same expression should occur in two propositions and should have exactly the same meaning in both cases. It must therefore have a meaning of its own, independent of the other parts of the proposition. In the case of incompletely defined concept words there is no such independence: what matters in such a case is whether the case at hand is the one to which the definition refers, and that depends on the other parts of the proposition. Such words cannot therefore be acknowledged to have an independent meaning at all. This is why I reject conditional definitions of signs for concepts.

The situation is quite similar for relations. A conditional definition of a sign for a relation, as for example for identity, decides only in some cases, not in all, whether the relation holds. Thus your definition I, sect. 4, 2 decides whether a is identical with b only in the case where a and b are classes; it does not therefore give the sign of identity a meaning independent of a and b; i.e., it does not give it a meaning at all. Besides the two cases

$$a \text{ is identical with } b$$

and

$$a \text{ is not identical with } b$$

there still remains a third case here, that of undecidability, whereas logic does not tolerate a third case.

Besides, I have still other doubts about individual definitions. Thus I cannot understand *Formulary* I, sect. 5, 2 as a definition:

$$\text{`}a, b \; \varepsilon \; \mathrm{K} \cdot \supset : f \; \varepsilon \; b/a \cdot x, y \; \varepsilon \; a \cdot x = y \cdot \supset \cdot fx = fy\text{'}$$

Where is the sign of identity on whose left side occurs the sign to be defined? Further, the function letter 'f' occurs here without an argument place in the combination '$f \; \varepsilon \; b/a$'.[3] We also have this case in I, sect. 5, 11 for the

---

3 Meaning: $f(x)$ is an element of b if x is an element of a.

function letters '*f*' and '*g*'. This is to misunderstand the essence of a function, which consists in its need for completion. One particular consequence of this is that every function sign must always carry with it one or more places which are to be taken by argument signs; and these argument places – not the argument signs themselves – are a necessary component part of the function sign. I find a further mistake in I, sect. 5, 11:

$$\text{'}a \ \varepsilon \ \mathrm{K} \cdot \supset :f, g \ \varepsilon \ b/a \supset \ \therefore \ f = g \cdot = :x \ \varepsilon \ a \cdot \supset_x \cdot fx = gx\text{'}$$

Here the letter '*a*' occurs by mistake on the right-hand side of the definitional equation, not on the left-hand side. Evidently '*f* = *g*' could mean the same as '*x* ε *a* · ⊃$_x$ · *fx* = *gx*' only if the meaning of the latter expression was independent of '*a*'.

But even disregarding this, you will have to admit that each of your definitions of the sign of identity is incomplete by itself, though you will perhaps want to claim completeness for their totality. In that case, this totality would also have to be presented as such externally. But the *Formulary* is arranged in such a way that we do not know, after any of these definitions, whether it is the last one, or whether there are more to come to complete the whole. Let us suppose that this mistake has been corrected and that the different definitions of the sign of identity have been combined into a self-contained whole; we should then have to ask whether this whole now represents a complete definition, whether the cases taken into consideration exhaust the totality of possibilities, but also whether we are not presented with double determinations in some cases. Now we can easily tell that this totality too is not complete; for nothing has been said about cases where the sign for an object other than a class occurs on the one side of the sign of identity, while a sign for a class occurs on the other side, and we are told just as little about cases where there occurs, say, a proposition on the left and a sign for a class or a proper name on the right. Of course you intend the equation to be false in these cases; but this cannot be gathered from your definitions.

The reasons why I cannot admit that you have defined the sign of identity by means of the signs ∩, ⊃, and − are mainly the following:

(1) every one of these definitions is incomplete;
(2) even taken together they are still incomplete, in so far as they do not decide in every case whether there is an identity;
(3) they explain the sign of identity by means of itself.

I have dwelt so long on this, not so much for the sake of the sign of identity, but rather for the sake of the principles of definition, which come into question in many other places as well. I find that on the whole you are not strict enough in this, and that for this reason your reductions of logical forms to simpler ones frequently lack the power of conviction.

Further, where is the definition which explains the sign 'Λ' by means of the signs ∩, ⊃, and −? In the *Formulary*, I first find this sign in I, sect. 3, 1, but this is not called a definition, and if it were, would also be a faulty one. In your *Notations of Mathematical Logic*, this sign is explained in two places, but only in words, and so the signs ∩, ⊃, and — do not occur at all in the explanations.[4] Here we must again take into consideration what I said about multiple definitions. But the situation here is somewhat different from before, in that there are no cases to be distinguished and neither of the two explanations is incomplete. There remain only two possibilities here: either these explanations contradict each other, or the content of the one is a consequence of the other, in which case the former would have to be proved as a theorem.

Besides, when it comes to listing all primitive signs, one should also mention the relational sign 'ε'; for it cannot be defined in terms of other signs. The dash above the function letter, used to designate conversion, must also count as a primitive designation. Among the primitive signs one must also count 'K', as well as many of the signs for classes listed in sect. 2 of your *Notations*. I have no need for a special primitive sign for this 'K', but can write for it

$$\grave{a}(\grave{\varepsilon}(-\varepsilon \cap a) = a),$$

which consists only of the signs I introduced earlier. Thus if I wanted to adopt the simple sign K and define it, I could do it by means of the equation

$$\grave{a}(\grave{\varepsilon}(-\varepsilon \cap a) = a) = \mathrm{K}.$$

And if I wanted to use the sign N with your meaning, I could reproduce your 'N ε K' as

$$\mathrm{N} \cap \grave{a}(\grave{\varepsilon}(-\varepsilon \cap a) = a)$$

or as

$$\grave{\varepsilon}(-\varepsilon \cap \mathrm{N}) = \mathrm{N}.$$

For these reasons I do not believe that your number of primitive signs is really smaller than mine. But I also do not regard the mere counting of primitive signs as an adequate basis for judging how profound the under-lying analysis is. I have, e.g., the sign ⊢, the judgement stroke, which serves to assert something as true. You have no corresponding sign, but you acknowledge the difference between the case where a thought is merely expressed without being put forward as true and the case where it is asserted. Now if the lack of such a sign in your conceptual notation had the effect that, upon close scrutiny, the number of your primitive signs turned out to be smaller, this would not allow us to conclude that your analysis was more profound; for the objective difference remains even if it is not mirrored in the signs. In making such a comparison, one would also have to take into consideration the number of independent stipulations and the things that can

---

4 Cf. G. Peano, *Notations de logique mathématique: Introduction au Formulaire de Mathématique* (Turin 1894), p. 7, line 4 from the end, and p. 11, line 5.

be done with them. Among such stipulations one would also have to count, e.g., that Ku is to mean the class of u (*Notations*, sect. 2), as well as some other propositions.

Thus the question whose underlying analysis is more profound does not seem to me to admit of a very simple answer. Various things would have to be considered: the number of original stipulations, the strictness of the principles of definition, and the number of things that can be done with the primitive signs. For the time being, therefore, I beg to doubt whether your analysis is more profound than mine.

There is still much to be said, e.g., about my use of Roman, German and Greek letters, a point on which you have misunderstood me, or about the conditions u, v ε K, which you think are missing from my proposition (32).[5] But this would take me further afield: I hope to write something about it on a later occasion.

                                                          Yours sincerely,
                                                          G. Frege

XIV/8 [xxxiv/8]   PEANO to FREGE   undated[1]

The great difficulties we have encountered in editing and printing F$_2$ §1 (sect. 1 of vol. II of the *Formulary of Mathematics*) have delayed the publication of No. 2 of vol. VI of our *Rivista* and hence also that of the extremely important letter from Mr Frege which will help to clarify several difficult and controversial points in mathematical logic. I will respond briefly to some of his observations.

In my review of Mr Frege's work I said that 'all the operations and relations between propositions and between classes reduce to three fundamental ones: ∩, ⊃, and —'; but undoubtedly, the system of primitive ideas we encounter in the whole of logic is more complex.

According to the collection of logical formulas recently published as F$_2$ §1, the designations which express in ordinary language the primitive ideas we could not express in symbols are nine in number, and expressed in P1 to 7, 70, and 100.

This reduction of primitive ideas does not yet amount to the last word. In the notes to F$_2$ §1 I hinted at other ways of arranging the ideas of logic; and it is possible that the discovery of new logical identities will lead to a further reduction.

Among the primitive ideas figures also the idea of definition (P7), which is indicated by the symbol = Df; this can be read as 'equals by definition'.

---

5 ⟨11⟩, p. 86.
XIV/8. 1 This is Peano's reply to Frege's letter of 29.9.1896 (XIV/7 above). It was published in *Rivista di matematica* 6 (1896–9), pp. 60ff, after Frege's letter, and later reprinted in G. Peano, *Opere Scelte* II (Rome 1958), pp. 295ff.

Although the signs = and Df are written separately, for in the *Formulary* the latter is always written at the end of the line, they nevertheless form a single symbol. Mr Burali-Forti in his *Mathematical Logic*, 1894, conjoins them and writes $=_{Def}$;[2] Mr Pieri fuses them typographically and writes $\equiv$;[3] but these are only formal differences.

Mr Frege wants any sign to have a single definition. And this is also my opinion when we are dealing with a sign which does not contain variable letters (F_2 §1 P7). But if what is being defined contains variable letters, that is, if it is a function of these letters, then I see, in general, the need for giving a conditional definition of that expression, or a definition *cum hypothesi* (ibid., P7'), and for giving as many definitions as there are kinds of things on which the operation in question is carried out. Thus the formula $a + b$ is defined the first time when a and b are whole numbers, then a second time when they are fractions, and then when they are irrational, or complex. The same sign + is found between infinite and transfinite numbers (F_1 VI), and it must then be given a new definition. It is found between two vectors and will then be defined anew, and so on. And with the progress of science the meaning of this formula is being extended further and further. The various meanings of the mark $a + b$ have common properties; but these are insufficient for specifying all the values that this expression may have.

The same happens with the formula $a = b$; in some cases its meaning can be assumed as a primitive idea; in others it is defined: in arithmetic in particular, the equality of integers being granted, equality is defined for rational numbers, for irrational ones, for imaginary numbers, etc. In geometry it is customary to define the equality of two areas, of two volumes, the equality of two vectors, etc. With the progress of science the necessity for extending the meaning of the formula $a = b$ makes itself increasingly felt. The various meanings of it have common properties; but I do not see that they are enough to specify all the possible meanings of equality.

Besides there is a great divergence of opinions among authors about the concept of equality; and a study of that question would be very useful, especially if it was conducted with the aid of symbols rather than in words.

Prop. F_1 1 §5 P2 is joined to the previous one and with it constitutes a definition. The definition reads:

'Let a and b be classes, and let us say that f is a b function of the a's; however, if we take x from class a, fx is a b; and if two individuals x and y, taken from class a, are identical, the corresponding values of the function are also identical.'

---

2 C. Burali-Forti, *Logica matematica* (Milan 1894). The mathematician Cesare Burali-Forti (1861–1931) was Peano's assistant from 1894 to 1896 and collaborated with him on the *Formulaire*.
3 Mario Pieri (1860–1913) taught at the time both at the Military Academy and at the University of Turin (like Peano himself) and was chiefly concerned with projective and descriptive geometry.

In the new collection of logical formulas, it was possible to divide this definition into two propositions, F_2 §1 P500 and P503, and this was possible because of the new P80.

The observation that F_1 §5 P11 is not homogeneous is correct, and this proposition is no longer found in F_2.

The *Formulary of Mathematics* is not the work of an individual, but is turning more and more into a collaborative effort; and all observations that contribute to its growth and perfection will be received with gratitude.

G. Peano

XIV/9 [xxxiv/9]   PEANO to FREGE   3.10.1896

Turin
3 October 1896

Dear Colleague,

After a long trip I found your letter of 29 September. I am sending you the printer's proofs of the beginnings of vol. II of the *Formulary* where there are some questions concerning you.[1] I shall write you at greater length in a few days. I have enjoyed reading your learned observations.

Yours sincerely,
G. Peano

XIV/10 [xxxiv/10]   PEANO to FREGE   14.10.1896

Turin
14 October 1896

Dear Colleague,

Thank you for the review you published in the Göttingen proceedings[1] and for your postcard. I should very much like to discuss so many interesting questions with you; but my work, notably the continued publication of the *Formulary*, prevents me from doing it at present. I shall touch on only a few points.

I propose to read the signs ε, $\supset$, $\cap$, $\cup$ . . . as 'is, is contained, and, or, . . .' so that it will be possible to *read* the formulas (*Introduction*, sect. 6);[2] but this is only a matter of pronunciation; it is not permissible to always substitute the signs ε, $\cap$, $\cup$ for the words *is*, *and*, *or* of ordinary language, or to attribute to these signs all the properties of these words.

XIV/9. 1 G. Peano, *Formulaire de mathématiques* II, sect. 1 (Turin 1897).
XIV/10. 1 The reference is presumably to ⟨15⟩. This is confirmed by the discussion of examples like $(3 > 2) \supset (7^2 = 0)$.
2 G. Peano, *Notations de logique mathématique*: *Introduction au Formulaire de Mathématique* (Turin 1894).

Consequently, in sect. 32 I read the formula $y\alpha|x$ as *y is an α inverse of x*; the sign of inversion | is written after the sign of relation α. If I want to separate these signs by parentheses, I write $y(\alpha|)x$.

In sect. 33 the formula $x\alpha\mathsf{t}v$ represents a relation between x and v derived from relation α by means of a suitable definition. If I want to put parentheses, I write $x(\alpha\mathsf{t})v$; $\alpha\mathsf{t}$ is the new sign of relation. This formula does not decompose into $x\alpha(\mathsf{t}v)$, which has no menaing. I propose to read the sign t as *each*; but the formula $x\alpha\mathsf{t}v$ always represents *the relation α each between x and v* and not *the relation α between x and each v*.

I have given two names to the sign $\supset$: '*we deduce*' and '*is contained*', and it can also be read in many other ways. This does not mean that the sign $\supset$ has several meanings. My idea is better expressed by saying that the sign $\supset$ has a single meaning, but that in ordinary language this meaning is represented by several different words, according to the circumstances. Similarly for the sign Λ.

Formulas such as

$$(3 < 2) \supset (7^2 = 0)$$
$$(7^2 = 0) = \Lambda$$

are not to be encountered in the *Formulary*. I mentioned them in the *Introduction* because they might appear in some demonstration.[3] But one could also conduct the demonstrations in such a way as to never encounter them. The sign $\supset$ is always written between *conditional* propositions; it always has the subscripts $x, y, \ldots$, either written or understood, except for the use that can be made of it according to *definition 5* of the *printer's proofs*,[4] a use which could be suppressed by not making use of this definition.

I have reread your books *Conceptual Notation* and *Basic Laws* with renewed pleasure,[5] for I understand them better and better. But I still find some obscure points.

Prop. 2, p. 26 of *Conceptual Notation* coincides with P29 of *Formulary* II, and I shall attach your name to it.[6]

Your prop. 1 coincides with P23 of the *Formulary*.
———— 5 ———— P26 ————
———— 28 ———— P77 ————
———— 29 ———— P74' ————
————31 and 41———— P73 ————
———— 36 ———— P80 ————

3 Cf. e.g. sect. 15.
4 The reference is evidently to the draft of *Formulaire* II which Peano sent to Frege for his approval (cf. XIV/9 above). The draft seems to differ substantially from the final version (cf. note 7 below).
5 $= \langle 1 \rangle$ and $\langle 11 \rangle$.
6 Frege has $(c \to (b \to a)) \to ((c \to b) \to (c \to a))$, whereas Peano has $(a \wedge b \to c) \wedge (a \to b) \to (a \to c)$ – both in a more modern notation. Peano substitutes conjunction for implication also in other cases. Incidentally, Peano did mention Frege in connection with P29 of *Formulaire* II.

Is your prop. 52 identical with my P64?

$$c = d \cdot c \ \varepsilon f \cdot \supset \cdot d \ \varepsilon f$$

(I shall add $f \varepsilon$ K.)
Your 54 coincides with my 61,
—— 55——————————— 62.
Is the proposition that follows your 58 more or less my 25?
——————————————————— 62 ——————— 25?

$$(g, f \varepsilon \text{ K}) \cdot x \ \varepsilon \ g : g \supset f \cdot \supset \cdot x \ \varepsilon \ f$$

Your 65 = my P26.[7]

As you so well put it, my principal aim is to publish the *Formulary* and not to deal exclusively with logic or with a particular subject. My work is to cut up the old *Formulary* with scissors and to put the pieces together again

---

7 In the following table, Frege's and Peano's versions have been placed side by side. Peano is referring to his draft; but this appears to have been revised substantially, so that the corresponding formulas appear in *Formulaire* II under different numbers. In the case of Frege's formulas, the assertion sign '⊢' has been dispensed with. In the case of Peano's formulas, premises like '$a \varepsilon$ K' have been omitted; moreover, '=' is occasionally reproduced as '↔', and '⊃' as '⊂'.

Frege (*Begriffsschrift*)		Peano		
No.	Formula	No. (draft)	No. (final)	Formula
1	$a \to (b \to a)$	P23	P23	$a \wedge b \to a$
5	$(b \to a) \to ((c \to b) \to (c \to a))$	P26	P26	$(a \to b) \wedge (b \to c) \to (a \to c)$
28	$(b \to a) \to (\neg a \to \neg b)$	P77	P111	$(a \to b) \to (\neg b \to \neg a)$
29	$(c \to (b \to a)) \to (c \to (\neg a \to \neg b))$	P74'	P108	$(a \wedge b \to c) \to (a \wedge \neg c \to \neg b)$
31	$\neg \neg a \to a$	P73	P106	$\neg \neg a \leftrightarrow a$
41	$a \to \neg \neg a$			
36	$a \to (\neg a \to b)$	P80	P114	$a \wedge \neg a \to b$
52	$c = d \to (f(c) \to f(d))$	P64	P80	$x = y \wedge (x \varepsilon a \to_a y \varepsilon a)$
54	$c = c$	P61	P41	$a = a$
55	$c = d \to d = c$	P62	P45	$a = b \to b = a$
flwg.58	$\wedge_a(g(a) \to f(a)) \to (g(b) \to f(b))$	P25	P25	$a \subset b \wedge x \varepsilon q \to x \varepsilon b$
flwg.62	$g(x) \to (\wedge_a(g(a) \to f(a)) \to f(x))$	P25	P25	
65	$\wedge_a(h(a) \to g(a))$ $\to (\wedge_a((g(a) \to f(a))$ $\to (h(x) \to f(x))))$	P26	P26	$a \subset b \wedge b \subset c \to a \subset c$

Note: In the case of P25 and P26, the analogies become clearer if one notes that Peano explains $a \supset b$ as $\wedge_x (x \varepsilon a \to x \varepsilon b)$. Peano's '⊃' has therefore been translated as '⊂' in these cases, whereas on the second line '⊃' has been translated as '→'.

with paste, while inserting all the numerous additions and corrections sent to me by my collaborators. Unfortunately I therefore do not have the time to examine every point; and I cannot derive as much profit from your books as would be desirable. You can correct the printer's proofs yourself, put your name to all the formulas for which you claim priority, add the new formulas you have found, and thus complete this part of the *Formulary*.[8]

From the printer's proofs you can also see that the number of primitive ideas, in terms of which all the others are expressed, is relatively high. There are around 10 from pure logic, 2 from arithmetic, etc. Among the primitive ideas I consider those represented by the signs ε, K, etc. In my review in volume V of the *Rivista*, p. 125, I meant to say that all the relations between propositions can be expressed by the signs $\cap$, $\sim$, and $\supset$. But one must, of course, add the sign ε to represent a relation between an individual and a class, etc.

I do in fact give several definitions of the equality $a = b$; sometimes this relation is regarded as primitive. In the formula

$$a, b \, \varepsilon \, \mathrm{K} \cdot \supset : a = b \cdot = \cdot a \supset b \cdot b \supset a \qquad \text{Def.}^9$$

the sign $\cdot = \cdot$ is joined to the sign *Def*; one could write

$$\underline{\hspace{3cm}} \cdot \quad : \underline{\hspace{1.5cm}} \cdot =_{\text{Def}} \underline{\hspace{3cm}}$$
'. . . . . means by definition . . . . .'

and this defines a new meaning, or a new case of meaning, for the sign $=$. Subsequently the sign $=$ is given constantly new meanings; equality is defined between two rational numbers, between two irrational ones, between two complex numbers, between two vectors, between two quaternions, etc. All these definitions contain an hypothesis expressing the meaning of the variable letters; for there will be no end to definitions of equalities. In the same way, $a + b$ is first defined when a and b are N (*Formulary* II, § 2, P21),[10] and then when they are n (ibid., P115);[11] after that when they are rational numbers, irrational ones, vectors, quaternions, and there is no end to definitions of additions.

But I believe that all this variety reduces to a simple question of notation. I have classified the definitions of the Formulary into four species:

(1) Definition of a sign (Introduction, sect. 36). There is no hypothesis.
(2) Definition of an expression containing variable letters (ibid., sect. 37). There is always an hypothesis.

---

8 In *Formulaire* II, Frege is mentioned explicitly only in connection with P29 (cf. note 6) and P121, as well as in the *Notes* on pp. 36 and 46.
9 This formula appears as P16 in *Formulaire* II.
10 In the final edition of *Formulaire* II, the definition of addition for natural numbers is to be found in P011.
11 In the final edition of *Formulaire* II, the definition of addition for whole numbers is to be found in P034.

(3) Definition by abstraction (ibid., sect. 38).

(4) Finally there are ideas, which I call primitive ideas, which are determined or, if you like, defined by means of their properties (ibid., sect. 42).

These distinctions emerge if one makes use of symbols, or rather, of my symbols. If one adopted other symbols, a different classification might perhaps emerge.

According to me, a proposition *which has meaning* is either true or false. But it is necessary that it have a meaning. For one can write propositions without meaning. For example,

$$a, b \ \varepsilon \, points \cdot \supset \cdot (a + b)^2 = a^2 + 2ab + b^2$$

has no meaning in an ordinary classroom situation. It has a meaning, and is correct, if the multiplication is taken to be commutative. It has a meaning, but is false, if the multiplication is not commutative.

In *Formulary* I, sect. 5, propositions 1 and 2 together constitute a single definition, written – not quite properly – on two lines:[12]

$$a, b \ \varepsilon \ K \, . \supset \, . \, . \ f \ \varepsilon \ b\, \mathfrak{J} \, a \, . = \, : x \, \varepsilon \, a \, . \supset_x . \ f \, x \, \varepsilon \, b :$$
$$x, y \ \varepsilon \ a \, . \ x = y \, . \supset . \ \mathfrak{f} x = \mathfrak{f} y$$

This should be read as follows:

'Let *a* and *b* be classes. We shall say that *f* is the correspondence of the *a*'s to the *b*'s if the following conditions are satisfied:

(1) If one writes the sign *f* before an arbitrary individual *x* of class *a*, one obtains an individual *fx* of class *b*.

(2) And if, in operating on two identical individuals *x* and *y* of class *a*, the resulting *fx* and *fy* are also identical.'

But now, in *Formulary* II, I think it better to separate the two propositions, by presenting one (P18 and P18′) as a definition and the other (P23) as a primitive proposition, for it expresses a property of the sign =.[13]

All the propositions about functions in sect. 5, vol. 1, part I of the *Formulary* which are no longer to be found in vol. II present difficulties or irregularities; I shall explain the reason for this in the *Rivista*.[14]

---

12 At the place cited by Peano we find '*b|a*' instead of '*b* ℐ *a*'. The latter designation is to be found in formula 500 of *Formulaire* II. In formula 500 the definition has, incidentally, been reduced to a single line, and the second line has been replaced by P503. ['ℐ' has the same meaning as '/', for which see note 3 to XIV/7.]

13 In the final version of *Formulaire* II, the propositions in question are P500, P501 and P503.

14 G. Peano, 'Sulle formule di logica', *Rivista di matematica* 6 (1896–9), pp. 48–52.

I shall publish your letter of 29.9.96 in the 2nd fascicle of the *Rivista*,[15] which will appear some months from now together with part of volume II of the *Formulary*. But you would be doing something very useful and very pleasant for the readers of the journal if you yourself were to explain the whole of your notation and to compare it with the notation of the *Formulary* with which our readers are already familiar. I believe that a classification of your notation into primitive and derived signs, with some examples, could be contained within a few pages and would not entail much work for you. There is still a large number of questions to be dealt with. For example, in my *Principles of Arithmetic*[16] and in the *Formulary*, vol. II, I take N and + as primitive ideas. In your book you define them. I shall be very happy to publish in the *Formulary* all your formulas and all your definitions and demonstrations, provided that they can be composed by using ordinary typographical characters.

The *Rivista di matematica* is only concerned with the *Formulary* and everything that concerns the latter. Every contribution must contain something concerning symbols.

Have you received my *Principles of Arithmetic* of 1889? I should like to thank you for your collaboration, and I shall send you some of my works and the volumes of the journal as tokens of my esteem.

Yours sincerely,
G. Peano

XIV/11 [xxxiv/11]   FREGE to PEANO   undated

The importance of the matter prompts me to say a few more words about multiple and conditional definitions. You agree with me on the definitions of proper names, but not on those of function signs. Yet you do not refute my reasons, but appeal to historical and practical considerations. It is true that as a matter of historical fact the meaning of a sign, e.g., that of the addition sign, has been extended step by step; but it is objectionable to retain this procedure in a systematic presentation. As I explained before, logic requires thoroughly fixed and definite limits. The fluctuations between limits and the rudimentary concepts and relations which are introduced by such extensions can only endanger the certainty of its inferences. You actually admit this; for you say nothing to the contrary; but practical reasons prevent you from acknowledging my requirement because you do not know how it can be satisfied. This is indeed a problem which is not very easily solved, but it must be solved, for logical requirements must not be suppressed because of practical difficulties. Now as far as the arithmetical signs for addition,

---

15 Cf. XIV/7 above.
16 *Arithmetices principia, nova methodo exposita* (Turin 1889).

multiplication, etc. are concerned, I believe we shall have to take the domain of common complex numbers as our basis; for after including these complex numbers we reach the natural end of the domain of numbers. Accordingly, I think it right to state in a single definition what meaning the combination of signs '$a + b$' is supposed to contain when any proper names are substituted for a and b. This meaning must of course agree with the one generally assumed when a and b are replaced by signs for common complex numbers (under which I include real numbers). It makes relatively little difference what the resultant meaning is when a or b or both are replaced by names of objects which are not common complex numbers, provided only that there is always a meaning there. This would fix the meaning of the addition sign once and for all; we should reach an end and gain a firm basis on which to construct our inferences with certainty. We should of course be giving up the possibility of making suitable determinations afterwards for higher complex numbers, vectors, etc., and we might perhaps find it necessary to introduce different signs for these cases. But we ought never to let such minor inconveniences deter us from satisfying the inescapable requirements of logic. As far as the equals sign is concerned, your remark that different authors have different opinions about its meaning leads to considerations that are both surprising and not very favourable to mathematics. If we consider that very many mathematical propositions present themselves as equations and that others at least contain equations, and if we place this against your remark, we get the result that mathematicians agree indeed on the external form of their propositions but not on the thoughts they attach to them, and these are surely what is essential. What one mathematician proves is not the same as what another understands by the same sign. We only seem to have a large common store of mathematical truths. This is surely an intolerable situation which must be ended as quickly as possible. As far as I am concerned, I take identity, complete coincidence, to be the meaning of the equals sign, and in definitions at least there seems to me to be no other possible meaning. What stands in the way of a general acceptance of this view is frequently the following objection: it is thought that the whole content of arithmetic would then reduce to the principle of identity, $a = a$, and that there would be nothing more than boring instances of this boring principle. If this were true, mathematics would indeed have a very meagre content. But the situation is surely somewhat different. When early astronomers recognized that the evening star (Hesperus) was identical with the morning star (Lucifer), or when an astronomer now brings out by calculation that a comet he observed is identical with one mentioned in ancient reports, this recognition is incomparably more valuable than a mere instance of the principle of identity – that every object is identical with itself – even though it too is no more than a recognition of an identity. Accordingly, if the proposition that $233 + 798$ is the same number as 1031 has greater cognitive value than an instance of the principle of identity, this does not prevent us from taking the equals sign in '$233 + 798 = 1031$' as a sign of identity. What a chaos of numbers we should have otherwise! There would not be a single number which was the first prime number after 5, but infinitely many: 7, 8 − 1, (8 + 6):2, etc. We should not speak of 'the sum of 7 and 5' with the definite article, but of 'a sum' or 'all sums', 'some sums', etc.; and hence we should not say 'the sum of 7 and 5 is divisible by 3' but 'all sums of 7 and 5 are divisible by 3'. How, then, does 3 · 4 differ from 9 + 3? In nothing but the signs. There is no property which (3 · 4) has but (9 + 3) lacks, and conversely. Of course, we must then not place the equals sign between signs for equally long lines or equally large areas. But we can nevertheless make use of the equals sign even in these cases. Instead of putting the signs for the lines on both sides, we can put the signs for the numbers we get by measuring these lines using the same unit. Or we can proceed as follows: every line determines a class of equally long lines; if two lines are equally long, the two classes coincide; and this again gives us an identity. Let A and B be lines and let '$A \cong B$' say that these lines are congruent. Then, in my notation, $\acute{\varepsilon}(\varepsilon \cong A)$ is the class of lines equal in length to A, and

$$\acute{\varepsilon}(\varepsilon \cong A) = \acute{\varepsilon}(\varepsilon \cong B)$$

is the essential content of '$A \cong B$' in the form of an equation. In similar cases, too, we can proceed in this way and express agreement in a certain respect in the form of an equation, without denying the meaning of the equals sign given above. Of course, this does not yet explain how it is possible that identity should have a higher cognitive value than a mere instance of the principle of identity. In the proposition, 'The evening star is the same as the evening star' we have only the latter; but in the proposition, 'The evening star is the same as the morning star' we have something more.

How can the substitution of one proper name for another designating exactly the same heavenly body effect such changes? One would think that it could only affect the form and not the content. And yet anyone can see that the thought of the second proposition is different, and in particular that it is essentially richer in content than that of the former. This would not be possible if the difference between the two propositions resided only in the names 'evening star' and 'morning star', without a difference in content being somehow connected with it. Now both names designate the same heavenly body; they have, as I put it, the same meaning; so the difference cannot lie in this. At this point my distinction between sense and meaning comes in in an illuminating way. I say that the two names have the same meaning but not the same sense, and that is shown by this, that the speaker need not know anything about the agreement in meaning, as most people ignorant of astronomy will not, in fact, know anything about it; but the speaker will have to connect a sense with the name unless he is babbling senselessly. And the sense of the name 'morning star' is indeed different from that of the word 'evening star'. And so it happens that the thought of our first proposition is different from that of the second; for the thought we express in a pròposition is the sense of the proposition. This is probably one of the cases of which you say, in your notice of my *Foundations of Arithmetic* (*Rivista di matematica*, vol. V, p. 127), that according to the dictionaries two of the words I use in different ways correspond to the same Italian word. The Italian *senso* seems to me to come closest to the word 'sense' (*Sinn*), and the Italian *significazione* to the word 'meaning' (*Bedeutung*). Dictionaries generally mix up the two by citing as corresponding words not only those that have the same sense but also those that have merely the same meaning. But only in the former case do we have an exact correspondence. Thus I find, e.g., for 'evening star': *Venere* and *Espero*, and for 'morning star': *Venere* and *Lucifero*. Accordingly, if we wanted to translate the sentence, 'The morning star is identical with the evening star' as, *La Venere è identica colla Venere*, we should be missing the sense (the thought). On the other hand, there is no objection to translating it as, *Il Lucifero è identico col Espero*. *Venere* has indeed the same *significazione* as 'morning star' but not the same *senso*. In the same way, the sense of '5 + 2' is also different from the sense of the combination of expressions '4 + 3', and consequently, the thought we express in the formula '$2^4 + 3 > 2 + 3$' is also different from the sense of the formula '5 + 2 > 5'. What we are talking about when we use a name in a proposition is not the sense of the name but its meaning. If we use the expression 'the sun' in a proposition, we mean by it a heavenly body which is in outer space and which is known to have mass.* But the sense of the word 'sun' is not somewhere in space, nor does it have mass. Thus if I write: '5 + 2 = 4 + 3', I do not mean by it that '5 + 2' and '4 + 3' have the same sense, but that they have the same meaning, that they designate the same number.** So nothing stands in the way of my using the equals sign as a sign of identity. But once this has been established, it is no longer possible to give definitions for cases where this sign occurs between rational, irrational or imaginary numbers, as you would have it, but from the established meaning of the equals sign, together with the explanation of the signs between which it occurs, it must follow by itself whether the equation is true. There are also other strong logical objections to your procedure. For it so happens that one defines neither the meaning of the equals sign by itself nor, for example, an irrational number by itself, but that one explains an expression in which both occur at the same time by stating when irrational numbers are supposed to be identical with one another. One then imagines one has explained both, whereas in truth one has explained neither; for from the meaning of a combination of signs one

---

* If I want to speak of words or signs themselves, not of their meanings, I place them in quotation marks.
** I have discussed sense and meaning in detail in *Zeitschrift für Philosophie und philosophische Kritik*, vol. 100.[1]

---

cannot yet infer the meanings of its individual component parts, not even if some of these signs should already be known. It is regrettable that there exists no agreement among mathematicians about the principles to be followed in defining. To produce such an agreement would be a worthwhile task for a mathematical congress. Complete lawlessness now prevails in this area, which is indeed convenient for mathematical writers but damaging to their science. There is not even agreement about what defining really is. Some believe they can create something by definition; but they do not discuss the barriers to this creative power nor the principles to be followed in defining creatively. And yet there must be such barriers; for no one thinks himself capable of creating an object with mutually contradictory properties. But neither does anyone prove, before defining creatively, that there is no contradiction, probably in the belief that any contradiction would have to jump to the eyes. How easy it would be to conduct all proofs if this was true! In sect. 33 of the first volume of my *Basic Laws of Arithmetic* I have laid down such principles,[2] and while everyone would, I think, agree with them in theory, they are almost always denied in practice.

XIV/12 [xxxiv/12]   PEANO to FREGE   7.1.1903

Turin
7 January 1903

Dear Colleague,

I have just received your *Basic Laws of Arithmetic*, second volume,[1] for which I thank you warmly, and I am planning to study it. I should be pleased and it would make my task easier if you could also send me volume I: I read it at the University Library which had acquired it at my request. Otherwise I shall be obliged to give volume 2 also to the Library instead of proposing its acquisition. I am sending you my *Arithmetic*[2] and a little note as tokens of my esteem. I am very anxious to read your book, for the question it studies is of great interest to me and to many mathematicians. But to shorten the time it would be useful if you tried to translate your propositions into the symbols of the *Formulary*. I shall try to do the same for my own part. And from the parallel between the two systems of writing, Mathematical Logic = Conceptual Notation will have much to gain.

Yours sincerely,
G. Peano

I have published parts 1 and 2 of vol. 4 of the *Formulary*; part 3 will soon follow.[3] I shall send it to you all together. But if you like, I can send you the published part now.

---

2 = ⟨11⟩.
XIV/12. 1 = GGA II.
2 G. Peano, *Aritmetica generale e algebra elementare* (Turin 1902).
3 G. Peano, *Formulaire mathématique* IV (Turin 1902–3).

XV FREGE–RUSSELL

Editor's Introduction

Bertrand Russell (1872–1970) corresponded with Frege from 1902 to 1912, though most of the correspondence belongs to the years 1902–4. The correspondence opens with Russell's announcement of his discovery of what is now known as the Russell paradox, and most of it is concerned with various solutions proposed by Russell and rejected by Frege to that paradox. But the correspondence also touches on most of the central concepts in Frege's philosophy of language: sense and meaning, object and concept, truth and falsity, proposition, and class. Russell's discovery of the paradox came at a time when Frege's life's work was substantially complete: vol. II of his *Basic Laws* was about to be published. Russell's major works were yet to come: at the time of the discovery he was preparing his *Principles of Mathematics* for publication. All of Russell's letters to Frege were written in German. At least one letter, written in 1912, is now lost. Russell's first letter (XV/1 below) and Frege's famous reply to it (XV/2) have been previously translated into English (cf. Jean van Heijenoort, ed., *From Frege to Gödel: A Source Book in Mathematical Logic, 1879–1931* (Cambridge, Mass., 1967)). Of all of Frege's letters to Russell only the last (XV/20) has survived in the original. Russell kept it in his possession, judging it to be purely personal. The rest were sent to Scholz and are now known only from photocopies.

XV/1 [xxxvi/1]   RUSSELL to FREGE   16.6.1902

Friday's Hill
Haslemere
16 June 1902

Dear Colleague,

I have known your *Basic Laws of Arithmetic* for a year and a half, but only now have I been able to find the time for the thorough study I intend to devote to your writings. I find myself in full accord with you on all main points, especially in your rejection of any psychological element in logic and in the value you attach to a conceptual notation for the foundations of mathematics and of formal logic, which, incidentally, can hardly be distinguished. On many questions of detail, I find discussions, distinctions and definitions in your writings for which one looks in vain in other logicians. On functions in particular (sect. 9 of your *Conceptual Notation*) I have been led independently to the same views even in detail. I have encountered a difficulty only on one point. You assert (p. 17) that a function could also constitute the indefinite element. This is what I used to believe, but this view now seems to me dubious because of the following contradiction: Let *w* be the predicate of being a predicate which cannot be predicated of itself. Can *w* be predicated of itself? From either answer follows its contradictory. We must therefore conclude that *w* is not a predicate. Likewise, there is no class (as a whole) of those classes which, as wholes, are not members of themselves. From this I conclude that under certain circumstances a definable set does not form a whole.

I am in the process of completing a book on the principles of mathematics, and I should like to discuss your work in it in great detail. I already have your books, or I shall buy them soon; but I should be very grateful to you if you could send me offprints of your articles in various journals. But if this should not be possible, I shall get them from a library.

On the fundamental questions where symbols fail, the exact treatment of logic has remained very backward; I find that yours is the best treatment I know in our time; and this is why I have allowed myself to express my deep respect for you. It is very much to be regretted that you did not get around to publishing the second volume of your *Basic Laws*; but I hope that this will still be done.

Yours sincerely,
Bertrand Russell

The above contradiction can be expressed in Peano's notation as follows:

$$w = \text{cls} \cap x \vartheta \, (x \sim \varepsilon x) \,.\, \supset : w \,\varepsilon\, w \,.\, = \,.\, w \sim \varepsilon\, w \,.^1$$

I have written about this to Peano, but he still owes me a reply.

XV/2 [xxxvi/2] FREGE to RUSSELL 22.6.1902

Jena
22 June 1902

Dear Colleague,

Many thanks for your interesting letter of 16 June. I am glad that you agree with me in many things and that you intend to discuss my work in detail. In accordance with your wishes I am sending you the following offprints:

(1) 'Critical Elucidation etc.'
(2) 'On the Notation of Mr Peano etc.'
(3) 'On Concept and Object'
(4) 'On Sense and Meaning'

XV/1. 1 This formula says that if w is the class of x such that $x \notin x$, then $w \in w \leftrightarrow w \notin w$. Russell takes over the notation essentially unchanged from G. Peano, *Formulaire de mathématiques* II, sect. 2 (*Arithmétique*) (Turin 1898); cf. formula 450 on p. vii:

$$u \,\varepsilon\, \text{Cls} \cdot \supset \cdot \text{Cls } u = \text{Cls} \cap x \vartheta \,(x \supset a) = \text{'class of } u\text{' Df.}$$

'a' in this formula should read 'u'; cf. op. cit., sect. 1 (*Logique mathématique*) (Turin 1897), p. 15, formula 450, though the notation here differs from the one chosen in sect. 2 by the use of 'K' instead of 'Cls', and of '$\overline{x\varepsilon}$' instead of '$x\,\vartheta$'.

(5) 'On Formal Theories of Arithmetic'[1]

I have received an empty envelope addressed in what seems to be your handwriting. I suspect that you had the intention of sending me something, but that it got lost by accident. If this is the case, I thank you for your good intention. I am enclosing the front of the envelope.

When I now reread my *Conceptual Notation*, I find that I have changed my view on some points, as you will see if you compare it with my *Basic Laws of Arithmetic*. Please cross out the paragraph on p. 7 of my *Conceptual Notation* beginning with 'We can just as easily' because it contains a mistake which, incidentally, did not have any undesirable consequences for the rest of the contents of my little book.[2]

Your discovery of the contradiction has surprised me beyond words and, I should almost like to say, left me thunderstruck, because it has rocked the ground on which I meant to build arithmetic. It seems accordingly that the transformation of the generality of an identity into an identity of ranges of values (sect. 9 of my *Basic Laws*) is not always permissible, that my law V (sect. 20, p. 36) is false, and that my explanations in sect. 31 do not suffice to secure a meaning for my combinations of signs in all cases. I must give some further thought to the matter. It is all the more serious as the collapse of my law V seems to undermine not only the foundations of my arithmetic but the only possible foundations of arithmetic as such. And yet, I should think, it must be possible to set up conditions for the transformation of the generality of an identity into an identity of ranges of values so as to retain the essentials of my proofs. Your discovery is at any rate a very remarkable one, and it may perhaps lead to a great advance in logic, undesirable as it may seem at first sight.

Incidentally, the expression 'A predicate is predicated of itself' does not seem exact to me. A predicate is as a rule a first-level function which requires an object as argument and which cannot therefore have itself as argument (subject). Therefore I would rather say: 'A concept is predicated

---

XV/2. 1 These are in the order given: $\langle 13 \rangle, \langle 15 \rangle, \langle 10 \rangle, \langle 9 \rangle$ and $\langle 6 \rangle$.

2 The mistake is in the first sentence of that paragraph, where Frege explains that the formula

$$\vdash \begin{array}{l} \Gamma \\ A \\ B \end{array}$$

'denies the case in which B is affirmed, but A and Γ are denied' (BS, sect. 5, p. 7). The mistake had already been pointed out by Ernst Schröder on p. 88 of his review of BS in *Zeitschrift für Mathematik und Physik* 25 (1880), pp. 81–94. Schröder made the plausible conjecture that in transforming an expression in his conceptual notation which should have led to '*non(non(B et non A)) et non Γ*', Frege had inadvertently skipped the second negation sign. Husserl's remarks on this passage, as reported by I. Angelelli, are less clear; cf. Appendix II (*Husserls Anmerkungen zur 'Begriffsschrift'*) in Gottlob Frege, *Begriffsschrift und andere Aufsätze*, 2nd ed. (Hildesheim 1964) (= $\langle 30 \rangle$).

of its own extension'. If the function $\Phi(\xi)$ is a concept, I designate its extension (or the pertinent class) by '$\acute{\varepsilon}\Phi(\varepsilon)$' (though I now have some doubts about the justification for this). '$\Phi(\acute{\varepsilon}\Phi(\varepsilon))$' or '$\acute{\varepsilon}\Phi(\varepsilon) \cap \acute{\varepsilon}\Phi(\varepsilon)$' is then the predication of the concept $\Phi(\xi)$ of its own extension.

The second volume of my *Basic Laws* is to appear shortly. I shall have to give it an appendix where I will do justice to your discovery. If only I could find the right way of looking at it!

Yours sincerely,

G. Frege

XV/3 [xxxvi/3]   RUSSELL to FREGE   24.6.1902

Friday's Hill

Haslemere

24 June 1902

Dear Colleague,

Many thanks for your letter and for sending me your works. I am sending you again the things that got lost in the mail. I had already corrected the mistake on p. 7 of your *Conceptual Notation*; but as you say, it has remained entirely without any damaging consequences.

In my opinion, concepts can in general be varied, and the contradiction arises only if the argument itself is a function of the function, i.e., if function and argument cannot vary independently. In the function $\varphi(\acute{\varepsilon}\varphi(\varepsilon))$, φ is the only variable, and the argument $\acute{\varepsilon}\varphi(\varepsilon)$ is itself (according to the usual manner of expression) a function of φ. It seems that functions of the form $\varphi\{F(\varphi)\}$, where F is constant and φ variable, are certainly permitted for every value of φ, though dangerous where the extension is in question. I call them quadratic forms: one might almost be inclined to introduce the imaginary into logic on the model of the imaginary in arithmetic.[1] With such functions we get at once a saturated function if we give the value of φ; yet they are not first-level functions, nor do they have constant arguments. The function $\dashv\vdash \varphi(\varphi)$ generates a contradiction similar to that generated by $\dashv\vdash$ $\varphi(\acute{\varepsilon}\varphi(\varepsilon))$.

I was led to the contradiction in the following way. As you, of course, know, Cantor proved that there is no greatest number. His proof is as follows:

$R\varepsilon 1 \rightarrow 1. \ \breve{\varrho} \supset \text{Cls'}\varrho.w = \varrho \cap x\mathfrak{z}(x \sim \varepsilon \mathfrak{1} \breve{\varrho}x).$

$$\supset_R.w \sim \varepsilon \varrho :\supset. \ \text{Nc' Cls'} \varrho \succ \text{Nc'}\varrho *$$

---

* These symbols are explained in *Revue de mathématiques* VII, 2.[2]

XV/3. 1 Cf. B. Russell, *The Principles of Mathematics* (Cambridge 1903; 2nd ed. London 1937), pp. 104, 107, 512 and 514.

2 The reference is presumably to B. Russell, 'Sur la logique des relations avec des applications à la théorie des séries', *Rivista di matematica* (= *Revue de mathématiques*) 7 (1900–1), pp. 115–48. However, not all the symbols used here are explained there.

(This is only the most essential part of the proof.)[3] Now there are concepts whose extension comprises everything; these should therefore have the greatest number. I tried to set up a one-one relation between all objects and all classes; when I applied Cantor's proof with my special relation, I found that the class Cls ∩ x з ($x \sim \varepsilon x$) was left over, even though all classes had already been enumerated. I have already been thinking about this contradiction for a year; I believe the only solution is that function and argument must be able to vary independently.

From what you say on p. 37, that a function name can never take the place of a proper name (I am speaking of the *Basic Laws*), there arises a philosophical difficulty. I know very well what good reasons there are to be found for this view; yet it is self-contradictory. For 'ξ can never take the place of a proper name' is a false proposition if ξ is a proper name, but otherwise it is not a proposition at all. If there can be something which is not an object, then this fact cannot be stated without contradiction; for in the statement, the something in question becomes an object. It therefore seems to me doubtful whether the φ in φx can be regarded as anything at all. But at this point we are plunging into philosophical logic.

On p. 49 you say that $\Gamma = \Delta$ has a meaning if Γ and Δ are proper names for ranges of values or names for truth-values. Yet on the preceding pages I find no explanation of $\Gamma = \Delta$ for the case where the one is a name for a range of values and the other the name of a truth-value, except for the case where

---

3 As a sketch of Cantor's proof, Russell's formula is hardly intelligible by itself. However, if it is compared with Russell's execution of the proof in B. Russell, 'On Some Difficulties in the Theory of Transfinite Numbers', *Proceedings of the London Mathematical Society*, series 2, vol. 4 (1907), part 1 (issued March 7, 1906), pp. 29–53, esp. p. 32, as well as with Russell's sketch of the proof in his *Principles*, sect. 349, it is possible to reconstruct the content of the formula as follows:

According to Peano's notation and Russell's use of the corresponding letters in the essay cited above and in the *Principles*, ρ is the domain and $\breve{\rho}$ the converse domain of the one-one relation R; Cls'ρ is the class of subclasses of ρ; Nc'ρ is the cardinal number of ρ; and the sign '⊃' between signs for classes is used, as in Peano, as the sign of inclusion, i.e., like '⊆' nowadays. The formula then says that the cardinal number of a class of subclasses of class ρ is greater than the cardinal number of ρ itself, because in every one-to-one correlation R of ρ with a class of subclasses of ρ (and in particular, with the class of *all* subclasses of ρ) the class w of all the elements of ρ which are not elements of the domain of R does not appear as an element of the domain of R. For if w appeared as the correlate of element x of ρ, then on the one hand, the assumption $x \in w$ would lead via the defining condition of w to $x \notin Rx$ and hence, because of the assumption $w = Rx$, to $x \notin w$ (and thus to its own contradictory), while on the other hand, the assumption $x \notin w$ would first lead via $w = Rx$ to $x \notin Rx$ and then, together with the universally valid $x \in \rho$ and via the defining condition of w, to $x \in w$ (and thus again to the contradictory assumption), and hence, taking everything together, to a contradiction.

However, if this reconstruction is correct, then Russell's formula between the two implication signs should read '$w \sim \varepsilon \rho$' instead of '$w \sim \varepsilon \breve{\rho}$'.

the range of values in question comprises everything or nothing. But I believe I did not understand you correctly on this point.[4]

Up to now I have read only your *Conceptual Notation* and your *Basic Laws*: I shall study the other works presently.

Yours sincerely,
Bertrand Russell

XV/4 [xxxvi/4]   FREGE to RUSSELL   29.6.1902

Jena
29 June 1902

Dear Colleague,

I have received your letter of the 24th and your publications; thank you very much for them.

Concerning the contradiction you found, perhaps I do not quite understand what you say about it. It seems that you want to prohibit formulas like '$\varphi(\acute{\varepsilon}\varphi(\varepsilon))$' in order to avoid the contradiction. But if you admit a sign for the extension of a concept (a class) as a meaningful proper name and hence recognize a class as an object, then the class itself must either fall under the concept or not; *tertium non datur*. If you recognize the class of square roots of 2, then you cannot evade the question whether this class is a square root of 2. If it should appear that this question could be neither affirmed nor denied, this would show that the proper name '$\acute{\varepsilon}(\varepsilon^2 = 2)$' was meaningless. Or should one present ranges of values (extensions of concepts, numbers) as a special kind of object such that certain predicates could be neither ascribed to them nor denied of them? This too would surely run into major difficulties.

Concerning your doubts regarding my proposition that a function name can never take the place of a proper name, we must distinguish sharply between a name or sign and its meaning. When I use a proper name in a proposition, I am not talking about this proper name but about the object it designates. But it can happen that I want to talk about the name itself; I then enclose it within quotation marks. In order to bring out the unsaturatedness of the function name, let me leave the argument place empty for once. I can then say:

'( )· 3 + 4' is a function name

and

'( )· 3 + 4' can never take the place of a proper name.

---

4 Russell seems indeed to have misunderstood Frege's stipulation in sect. 10 of GGA I (p. 17) that the range of values $\acute{\varepsilon}(-\varepsilon)$ is the true and the range of values $\acute{\varepsilon}(\varepsilon = \frown_{\alpha}^{} a = a)$ the false, by taking the two ranges of values to comprise 'everything or nothing', that is, the former everything and the latter nothing. Frege corrects this misunderstanding in the following letter.

You are correct in saying:

'"ζ can never take the place of a proper name" is a false proposition if ζ is a proper name';

but you are not correct in continuing:

'but otherwise it is not a proposition at all'.

On the other hand, it is correct to say:

If 'ζ' is not a proper name, then 'ζ can never take the place of a proper name' is not a proposition.

Here '"( )·3 + 4"' – with two sets of quotation marks – takes the place of 'ζ'. While '( )·3 + 4' is a function name, '"( )·3 + "' is a proper name, and its meaning is the function name '( )·3 + 4'. In the proposition 'Something is an object', the word 'something' takes an argument place of the first kind and stands for a proper name. Thus whatever we put in place of 'something', we always get a true proposition; for a function name cannot take the place of 'something'. Here we find ourselves in a situation where the nature of language forces us to make use of imprecise expressions. The proposition 'A is a function' is such an expression: it is always imprecise; for 'A' stands for a proper name. The concept of a function must be a second-level concept, whereas in language it always appears as a first-level concept. While I am writing this, I am well aware of having again expressed myself imprecisely. Sometimes this is just unavoidable. All that matters is that we know we are doing it, and how it happens. In a conceptual notation we can introduce a precise expression for what we mean when we call something a function (of the first level with one argument), e.g.: '$\acute{\varepsilon}\varphi(\varepsilon)$'.[1] Accordingly, '$\acute{\varepsilon}(\varepsilon·3 + 4)$' would express precisely what is expressed imprecisely in the proposition '$\zeta·3 + 4$ is a function'. Whatever we now put in place of '$\varphi(\ )$', we always get a true proposition because we can only put in names of functions of the first level with one argument, for the argument place here is of the second kind. Just as in language we cannot properly speaking say of a function that it is not an object, so we cannot use language to say of an object, e.g. ♃,[2] that it is not a function. You are correct in thinking that a function cannot properly be treated as something; for, as I said before, the word 'something' stands for a proper name. Instead of using the imprecise expression 'ζ is a function', we can say: '"( )·3 + 4" is a function name'. We cannot properly say of a concept name that it means something; but we can say that it is not meaningless. It is clear that function signs or concept names are indispensable. But if we admit this, we must also admit that there

---

XV/4 1 By using the *spiritus asper* instead of the *spiritus lenis* which he customarily uses in this place, Frege apparently wants to distinguish the expressions in question from his names for ranges of values.

2 This is the sign of the planet Jupiter; cf. GGA II, p. 84, and PW p. 227.

are some that are not meaningless, even though, strictly speaking, the expression 'the meaning of a function name' must not be used.

Regarding the last of the points you touch on, I shall make the following remark: $\acute{\varepsilon}(-\!\!-\ \varepsilon)$ is a class comprising only a single object, namely the true, and $\acute{\varepsilon}(\varepsilon = \neg\cup\!\!\!\cdot\, \mathfrak{a} = \mathfrak{a})$ is a class comprising only a single object, namely the false. If Γ is neither the one class nor the other but some other range of values, then Γ is distinct from the true because it does not coincide with $\acute{\varepsilon}(-\!\!-\varepsilon)$ and likewise distinct from the false because it does not coincide with $\acute{\varepsilon}(\varepsilon = \neg\cup\!\!\!\cdot\, \mathfrak{a} = \mathfrak{a})$. Thus if $\varDelta$ is a truth-value, '$\Gamma = \varDelta$' means the false.

The title of your paper 'Is Position in Time etc.'[3] makes me suspect that you might be interested in an essay I once published in the *Zeitschrift für Philosophie und philosophische Kritik* on a similar question.[4] I cannot find an offprint of it any more, nor do I remember the title, but if you wish I can look for it. You are probably familiar with my short paper 'On the Numbers of Mr H. Schubert'.[5]

I have not yet found the time to study your papers, but I hope to do it soon.

Yours sincerely,
G. Frege

XV/5 [xxxvi/5]   RUSSELL to FREGE   10.7.1902

Trinity College
Cambridge
10 July 1902

Dear Colleague,

Concerning the contradiction, I did not express myself clearly enough. I believe that classes cannot always be admitted as proper names. A class consisting of more than one object is in the first place not *one* object but many. Now an ordinary class does form *one* whole; thus soldiers for example form an army. But this does not seem to me to be a necessity of thought, though it is essential if we want to use a class as a proper name. I believe I can therefore say without contradiction that certain classes (namely those defined by quadratic forms) are mere manifolds and do not form wholes at all. This is why there arise false propositions and even contradictions if they are regarded as units. I do not prohibit formulas like $\varphi\{\acute{\varepsilon}\varphi(\varepsilon)\}$ in cases where $\acute{\varepsilon}\varphi(\varepsilon)$ means a real object; but we must not regard φ in such an expression as a variable because we would then have to consider values of φ for which this is not the case. This view rests on the assertion

---

3 B. Russell, 'Is Position in Time and Space Absolute or Relative?', *Mind* 10 (1901), pp. 293–317.
4 Frege is thinking of his essay on the law of inertia ⟨8⟩.
5 = ⟨16⟩.

that ranges of values are really classes, in the sense in which a class consists of a sum of objects. And this seems to me necessary because $\grave{\varepsilon}\varphi(\varepsilon)$ and $\grave{\varepsilon}\psi(\varepsilon)$ are *identical* if φx and ψx are *equivalent* for all values of x.*

Concerning function names, there still seems to me to be a difficulty. If we leave aside names altogether and speak merely of what they mean, then we must admit that there is no proposition in which a function takes the place of a subject. But the proposition 'A function never takes the place of a subject' is self-contradictory; and it seems to me that this contradiction does not rest on a confusion of a name with what it means.

I am not familiar with your paper on the numbers of Mr Schubert.[2] I should easily be able to find the other paper, on space and time, in the journal.[3] I sent you my paper because it contains a sharp criticism of psychologism and Kantianism.

My book is already in the press: I shall discuss your work in an appendix because it is now too late to talk about it in detail in the text.[4] When I read your *Basic Laws* for the first time, I could not understand your conceptual notation; I succeeded only when I began to notice the gaps in Peano's notation. Unfortunately my book was already completed at the time.

Yours sincerely,
Bertrand Russell

---

* What you say about the null class seems to me quite right, and in this case I must take a roundabout way.[1]

XV/6 [xxxvi/6]   RUSSELL to FREGE   24.7.1902

Trinity College
Cambridge
24 July 1902

Dear Colleague,

There is a difficulty in the theory of ranges of values with which I have been preoccupied for a long time and for which I find no satisfactory answer in your writings. Perhaps you can rid me of my doubts.

XV/5. 1 Russell does not elucidate this roundabout way. But cf. op. cit., sect. 73, pp. 73–6, where Russell likewise advocates an extensional concept of a class (as a 'sum of objects') and correlates a concept that has no object falling under it, not with a null class (for on the extensional conception there can be no such class), but with the class of all concepts 'identical' with the given empty concept, in the sense that the propositional functions representing all these concepts are logically equivalent.

2 $= \langle 16 \rangle$.

3 $= \langle 8 \rangle$.

4 Cf. Appendix A ('The Logical and Arithmetical Doctrines of Frege') in Russell, *Principles*, pp. 501–22.

If u and v are not ranges of values, we have

$$\underset{\textbf{a}}{\smile} \, a \cap u = a \cap v$$

because both $a \cap u$ and $a \cap v$ always mean the false. But from this we cannot infer $u = v$, for then any two objects would be identical as long as they were not ranges of values. It follows from this that u is not in general the range of values of the function $— a \cap u,$[1] but only if u is a range of values. From $\underset{\textbf{a}}{\smile} \, a \cap u = a \cap v$ we can only infer $u = v$ if we already know that u and v are ranges of values. But the question arises how this can be known. And in general, if one connects ranges of values closely with concepts, as you do, it seems doubtful whether two concepts with the same extension have the same range of values or only equivalent ranges of values. I find it hard to see what a class really is if it does not consist of objects but is nevertheless supposed to be the same for two concepts with the same extension. Yet I admit that the reason you adduce against the extensional view (*Archiv für systematische Philosophie* I, p. 444)[2] seems to be irrefutable.

Every day I understand less and less what is really meant by 'extension of a concept'. But in discussing your views, I should not like to bring up any unjustified objections, and this is why I permit myself to question you on the most difficult points.

It is easy to prove (as concerns the contradiction) that there is no one-one relation between all objects and all functions. For if φ_x is related to x, then $\overline{\top} \, \varphi_x(x)$ – for variable x – is a function which is not related by the relation under consideration to any x. This is why not all functions can be expressed in the form $x \cap u.$[3]

Yours sincerely,
Bertrand Russell

XV/7 [xxxvi/7]   FREGE to RUSSELL   28.7.1902

Forstweg 29
Jena
28 July 1902

Dear Colleague,

You write that a class consisting of more than one object is in the first place not one object but many. While an ordinary class forms a whole, certain classes do not form wholes but are mere manifolds, and what gives rise to contradictions is that they are nevertheless regarded as units. You do not mean to prohibit '$\varphi(\grave{\varepsilon}\varphi(\varepsilon))$' as such, but only when '$\grave{\varepsilon}\varphi(\varepsilon)$' does not mean

---

XV/6. 1 Russell evidently does not mean $—a \cap u$ but $a \cap u$. '$—$' appears to have been placed before '$a \cap u$' as an afterthought.
2 $= \langle 13 \rangle$.
3 Cf. Russell, *Principles*, sect. 102 (pp. 102ff) and sect. 348 (pp. 366ff).

a real object. In my opinion, you would have to prohibit '$\dot{\varepsilon}\varphi(\varepsilon)$' as such. If a class name is not meaningless, then, in my opinion, it means an object. In saying something about a manifold or set, we treat it as an object. A class name can appear as the subject of a singular proposition and therefore has the character of a proper name, e.g., 'the class of prime numbers comprises infinitely many objects'. We can distinguish the following cases:

(1) 'Socrates and Plato are philosophers'. Here we have two thoughts: *Socrates is a philosopher* and *Plato is a philosopher*, which are only strung together linguistically for the sake of convenience. Logically, *Socrates and Plato* is not to be conceived as the subject of which being a philosopher is predicated.

(2) 'Bunsen and Kirchoff laid the foundations of spectral analysis'. Here we must regard *Bunsen and Kirchhoff* as a whole. 'The Romans conquered Gaul' must be conceived in the same way. The Romans are here the Roman people, held together by customs, institutions and laws. An army is in this sense a whole, or system. We regard every physical body as a whole, or system, consisting of parts.

(3) 'The class of prime numbers comprises infinitely many objects'. Here the class of prime numbers is an object, but not a whole whose parts would be prime numbers. I would not say that this class consisted of prime numbers. This case differs from the preceding one as follows: first, a whole, a system, is held together by relations, and these are essential to it. An army is destroyed if what holds it together is dissolved, even if the individual soldiers remain alive. On the other hand, it makes no difference to a class what the relations are in which the objects that are members of it stand to one another. Secondly, if we are given a whole, it is not yet determined what we are to envisage as its parts. As parts of a regiment I can regard the battalions, the companies or the individual soldiers, and as parts of a sand pile, the grains of sand or the silicon and oxygen atoms. On the other hand, if we are given a class, it is determined what objects are members of it. The only members of the class of prime numbers are the prime numbers, but not the class of prime numbers of the form $4n + 1$, for this class is not a prime number. The only members of the class of companies of a given regiment are the companies, but not the individual soldiers. For wholes or systems we have the proposition that a part of a part is a part of the whole. This proposition does not hold for classes as regards the objects that are members of them. The relation of a company to a class of companies is quite different from the relation of this company to the regiment of which it is a part. The objects that are members of a class can at the same time form a system. But the system must still be distinguished from the class. The class of atoms that form the chair on which I am sitting is not the chair itself. A whole whose parts are material is itself material; on the other hand, I would not call a class a physical object but a logical one. It seems to me that you want to admit only systems and not classes. I myself was long reluctant to recognize ranges of values and hence classes; but I saw no other possibility

of placing arithmetic on a logical foundation. But the question is, How do we apprehend logical objects? And I have found no other answer to it than this, We apprehend them as extensions of concepts, or more generally, as ranges of values of functions. I have always been aware that there are difficulties connected with this, and your discovery of the contradiction has added to them; but what other way is there? In the notes at the end of sect. 1 of your essay 'On the Logic of Relations' you write: 'The cardinal number of a class *u* would be the class of classes similar to *u*'.[1] This agrees completely with my definition; but we must not then regard classes as systems; for the bearer of a number, as I have shown in my *Foundations of Arithmetic*,[2] is not a system, an aggregate, a whole consisting of parts, but a concept, for which we can substitute the extension of a concept. We can also try the following expedient, and I hinted at this in my *Foundations of Arithmetic*. If we have a relation $\Phi(\xi, \zeta)$ for which the following propositions hold: (1) from $\Phi(a, b)$ we can infer $\Phi(b, a)$, and (2) from $\Phi(a, b)$ and $\Phi(b, c)$ we can infer $\Phi(a, c)$; then this relation can be transformed into an equality (identity), and $\Phi(a, b)$ can be replaced by writing, e.g., '$\S a = \S b$'.[3] If the relation is, e.g., that of geometrical similarity, then '*a* is similar to *b*' can be replaced by saying 'the shape of *a* is the same as the shape of *b*'. This is perhaps what you call 'definition by abstraction'. But the difficulties here are not the same as in transforming the generality of an identity into an identity of ranges of values.

The difficulty in the proposition 'A function never takes the place of a subject' is only an apparent one, occasioned by the inexactness of the linguistic expression; for the words 'function' and 'concept' should properly speaking be rejected. Logically, they should be names of second-level functions; but they present themselves linguistically as names of first-level functions. It is therefore not surprising that we run into difficulties in using them. I have, I believe, dealt with this in my essay 'On Concept and Object'.[4] If we want to express ourselves precisely, our only option is to talk about words or signs. We can analyse the proposition '3 is a prime number' into '3' and 'is a prime number'. These parts are essentially different: the former complete in itself, the latter in need of completion. Likewise, we can analyse the proposition '4 is a square number' into '4' and 'is a square number'. Now it makes sense to fit together the complete part of the first proposition with that part of the second proposition which is in need of completion (that the

XV/7. 1 B. Russell, 'Sur la logique des relations avec des applications à la théorie des séries', *Rivista di matematica* 7 (1900–1), pp. 115–48.
2 GLA, part II, esp. sect. 23.
3 This tacitly presupposes that for any *a* there is a *b* such that $\Phi(a, b)$ or equivalently – by making use of symmetry – that for any *b* there is an *a* such that $\Phi(a, b)$; only then can reflexivity be deduced from symmetry and transitivity; and reflexivity, '$\Phi(a, b)$', is indispensable for abstracting from '$\Phi(a, b)$' to '$\S a = \S b$', for the new relation is supposed to be one of identity, and the identity $\S a = \S a$ must therefore hold.
4 Cf. $\langle 10 \rangle$, pp. 200ff [E.T. pp. 49ff].

proposition is false is a different matter); but it makes no sense to fit together the two complete parts; they will not hold together; and it makes just as little sense to put 'is a square number' in place of '3' in the first proposition. This difference between the signs must correspond to a difference in the realm of meanings; although it is not possible to speak of it without turning what is in need of completion into something complete and thus falsifying the real situation. We already do this when we speak of 'the meaning of "is a square number"'. Yet the words 'is a square number' are not meaningless. The analysis of the proposition corresponds to an analysis of the thought, and this in turn to something in the realm of meanings, and I should like to call this a primitive logical fact. This is precisely why no proper definition is possible here.

Yesterday I received your letter of the 24th. I am very pleased that you are so preoccupied with my writings and that you express your doubts to me. Unfortunately, I have neither the time nor the energy now to answer your letters as quickly as I should like to. Since I do not want to delay mailing this letter, I am postponing my reply to your last letter. Please do not let this keep you from putting all your doubts before me.

Yours sincerely,

G. Frege

XV/8 [xxxvi/8]　FREGE to RUSSELL　3.8.1902

Jena

3 August 1902

Dear Colleague,

Concerning the first question in your last letter,[1] the value of the function $\xi \frown \Delta$ is indeed always the same where Δ is not a range of values, and we cannot infer $u = v$ from

$$\dot{\varepsilon}\ a \frown u = a \frown v$$

Incidentally, the value of $\xi \frown \Delta$ where Δ is not a range of values is not the false but the extension of an empty concept. Nor is u in general the range of values of $\xi \frown u$. You ask how it can be known that something is a range of values. This is indeed a difficult point. Now, all objects of arithmetic are introduced as ranges of values. Whenever a new object to be considered is not introduced as a range of values, we must at once answer the question whether it is a range of values, and the answer is probably always no, since it would have been introduced as a range of values if it was one. You find it doubtful whether concepts with the same extension have the same range of values. Since for me the extension of a concept or a class is only a special case of a range of values, concepts always have the same range of values if they have the same extension; for the extension is the range of values. I can

XV/8. 1 Frege is referring to XV/6, which he received just before completing XV/7.

see indeed that, as a result of your discovery of the contradiction, propositions (1), (2), and their consequences are not true in all their generality as they were meant to be.[2] Indeed, it is still not clear to me in what way they are to be restricted. I wrote to you already in my last letter that what you would call a class is properly speaking a system, whole, or aggregate, and cannot replace what I call a class in the foundations of arithmetic and in logic. The truth is that properly speaking a system is not something logical.

You are correct in writing that not all functions can be designated by the form '$\xi \frown u$'; but the proof that there is no one-one relation between all objects and all functions strikes me as dubious. I believe that even the idea of objects standing in a one-one relation to functions is not quite clear. For standing in a one-one relation presupposes identity, and the relation of identity is a first-level relation which can hold only between objects and not between functions. If we speak of the identity of functions, we can only mean the identity of their ranges of values, or the identity of something connected one to one with ranges of values.[3]

Yours sincerely,
G. Frege

XV/9 [xxxvi/9]   Russell to Frege   8.8.1902

Little Buckland
Near Broadway, Worcs.
8 August 1902

Dear Colleague,

Many thanks for your explanations concerning ranges of values. I now understand the necessity of treating ranges of values not merely as aggregates of objects or as systems. But I still lack a direct intuition, a direct

---

2 Cf. GGA I, p. 75. Propositions (1) and (2) in Frege, i.e., '$f(a) = a \frown \grave{\varepsilon} f(\varepsilon)$' and '$f(a, b) = a \frown (b \frown \grave{\alpha} \grave{\varepsilon} f(\alpha, \varepsilon))$', correspond to the modern principles of comprehension, '$\wedge_x \cdot A(x) \leftrightarrow x \in_x A(x)$' and '$\wedge_x \wedge_y \cdot B(x, y) \leftrightarrow (x, y) \in_{x, y} B(x, y)$'.
3 In GGA I, p. 40, Frege defines the one-one relation for functions as follows:

$$\curlyvee\!\!\curlyvee\!\!\curlyvee\overset{c \quad b \quad a}{\underset{}{\,}}\!\!\begin{array}{l} b = a \\ \varphi(e, a) \\ \varphi(e, b) \end{array}$$

It is therefore a second-level function with argument places of the third kind, i.e., for first-level two-place functions. In GGA I, p. 55, it is replaced by the first-level function I, which is explained as follows:

$$\curlyvee\!\!\curlyvee\!\!\curlyvee\overset{c \quad b \quad a}{\underset{}{\,}}\!\!\begin{array}{l} b = a \\ e \frown (a \frown \xi) \\ e \frown (b \frown \xi). \end{array}$$

In both cases, the relation of identity between objects is presupposed. On the question of identity of concepts, cf. also PW pp. 121 and 182.

insight into what you call a range of values: logically it is necessary, but it remains for me a justified hypothesis.

The contradiction could be resolved with the help of the assumption that ranges of values are not objects of the ordinary kind; i.e., that $\varphi(x)$ needs to be completed (except in special circumstances) either by an object or by a range of values of objects or by a range of values of ranges of values, etc. This theory is analogous to your theory about functions of the first, second, etc. levels. In $x \frown u$ it would be necessary that u was a range of values of objects of the same degree as x; $x \frown x$ would therefore be nonsense. This view would also be useful in the theory of relations. You say, e.g. (*Foundations*, sect. 83), that every judgement $\varphi(a, b)$ expresses a relation between a and b. Take, e.g.,

$$R = S . \underset{\text{}}{\overline{\quad\quad}} (R)R(S)$$

i.e., R and S are identical, and the relation R does not hold between R and S. Let this equal $(R)T(S)$, where T is a relation. Given $R = T$ we then get the contradiction.[1] Therefore $\varphi(a, b)$ would not always express a relation between a and b. Yet it could be asserted that a relation

$$\underset{r=s}{\overline{}} \overset{r \frown (s \frown r)}{}$$

XV/9. 1 In Russell's notation, '$(R)\ T\ (S)$' says that relation T holds between R and S. In the more customary notation, '$T(R, S)$', Russell's reasoning proceeds as follows. He first defines:

$$T(R,S) \leftrightharpoons R = S \wedge \neg R(R,S).$$

By virtue of this definition, the biconditional holds between definiendum and definiens, and the substitution of T for R yields:

$$T(T,S) \leftrightarrow T = S \wedge \neg T(T,S).$$

Russell now employs the equation $T = S$ to reach the conclusion

$$T(T,T) \leftrightarrow \neg T(T,T),$$

where the implication involved in this step is evidently regarded as a material one. Russell also regards it as a formal implication in a later formulation of the paradox, where he defines at the outset

$$T(R,S) \leftrightharpoons \neg R(R,S),$$

whereupon he replaces both R and S by T so as to get the same contradiction as above. Cf. Russell, *Principles*, p. 521; 'On some Difficulties in the Theory of Transfinite Number and Order Types', *Proceedings of the London Mathematical Society*, series 2, vol. 4 (1907), part I (issued 7 March 1906); as well as 'Mathematical Logic as Based on the Theory of Types', *American Journal of Mathematics* 30 (1908), pp. 222–62, esp. p. 222.

between relations must be of a different logical *type* from a relation between objects, and $(R)R(S)$ would therefore be nonsense. $((R)R(S)$ is the same as $R \cap S \cap R.)$ For every function $\varphi(x)$ there would accordingly be not only a range of values but also a range of those values for which $\varphi(x)$ is decidable, or for which it has a sense. The striving for generality would accordingly be a mistake; i.e., $\neg \overset{a}{\smile} \neg \varphi(a)$ does not mean the assertion of $\varphi(x)$ for all values of x, but the assertion of all propositions of the form $\varphi(x)$.

Yours sincerely
Bertrand Russell

XV/10 [xxxvi/10]   FREGE to RUSSELL   23.9.1902

Jena
23 September 1902

Dear Colleague,

Forgive me if, due to various obstacles, I have only now got around to answering your last letter. I have considered various possible ways of resolving the contradiction, and among these also the one you indicated, namely that we are to conceive of ranges of values and hence also of classes as a special kind of objects whose names cannot appear in all argument places of the first kind. A class would not then be an object in the full sense of the word, but – so to speak – an improper object for which the law of excluded middle did not hold because there would be predicates that could be neither truly affirmed nor truly denied of it. Numbers would then be improper objects. We should also have to distinguish different argument places of the first kind, namely those that could take the names of both proper and improper objects, those that could only take the names of proper ones, and those that could only take the names of improper ones. The places on both sides of the equals sign would be of the first kind. In favour of this view it can be said that class names appear originally only on both sides of the equals sign (in my law V);[1] and prior to the introduction of a certain range of values, this does not tell us, for any previously-known function, what value it will take for this range of values as an argument. This lends support to the assertion that this range of values cannot be an argument for any previously-known function.

To avoid the case where the function $\xi = \zeta$ could have proper as well as improper objects as arguments, we might perhaps be inclined to assume a special kind of equality (identity) for improper objects; but there are major difficulties in this. If we now add the ranges of values of functions that can take both proper and improper objects as arguments, or those that only take

---

XV/10. 1 Cf. GGA I, p. 36, and further pp. 7, 14 and 16ff.

improper objects, etc., we get such a multiplicity of objects and functions that it becomes difficult to set up a complete system of logical laws. These doubts keep me for the time being from taking the way out you propose.[2]

I must confess that I do not quite understand what you write about relations. I must remark first of all that I distinguish between a relation and the domain of a relation. A relation is a function with two arguments whose value is always a truth-value; I call the double range of values of such a relation the domain of a relation. I should write your formula

$$R = S . \text{—T—} (R) R (S)^*)$$

as follows:

$$\text{T}\!\!\begin{array}{l} R \frown (S \frown R) \\ R = S \end{array}$$

Now you write: 'Let this equal $(R)T(s)$, where T is a relation. Given $R = T$ we get the contradiction'. I suspect that you propose to let

$$\grave{a}\grave{\varepsilon}\!\left(\text{T}\!\!\begin{array}{l} \varepsilon \frown (a \frown \varepsilon) \\ \varepsilon = a \end{array} \right) = T$$

We then have

$$\left(\text{T}\!\!\begin{array}{l} R \frown (S \frown R) \\ R = S \end{array} \right) = R \frown (S \frown T)$$

Now if we get the contradiction given $R = T$ (which, incidentally, is not evident to me), then this shows that R and T cannot truly be equated. Evidently, I do not quite understand your meaning here. A relation between relations is of a different logical type from one between objects. For the former is a second-level function, the latter a first-level one. But it may be asked whether relations between domains of relations are of a different logical type from relations between other objects, for domains of relations are objects if they are admissible at all. Here it may again be asked whether they are to be conceived as proper or improper objects. If they are admitted as improper objects, then they constitute a new kind of improper objects; and the multiplicity of kinds of objects and functions becomes even greater and hence also the difficulty of setting up a complete system of logical laws. But if domains of relations are conceived as proper objects, then a relation between domains of relations will be of the same logical type as one between objects in general.

---

* Here you use 'R' in a twofold manner, first as a two-sided function-sign, and second as the proper name of a relation, which I do not regard as admissible.

---

2 Cf. GGA II, pp. 254ff.

It may be asked whether there is not a characteristic mark by which those functions that have a range of values could be distinguished from those that have no range of values, and hence also, by which those concepts that have a class pertaining to them can be distinguished from those that have no extension; and here I am using the word 'class' for 'extension of a concept'.

Yours sincerely,

G. Frege

[On the last page of Frege's letter we find in Russell's handwriting:]

$$\mathbin{\mathsf{\tau}}\mathbin{\varphi}\!\!\left[\!\!\begin{array}{l}\varphi\,(\beta,\gamma) \\ \beta = \grave{a}\,\grave{\varepsilon}\,\varphi\,(a,\varepsilon) \\ \gamma = \grave{a}\,\grave{\varepsilon}\,\varphi\,(a,\varepsilon)\end{array}\right.$$

$$\sim (\beta = \grave{a}\,\grave{\varepsilon}\,\varphi\,(a,\varepsilon) \,.\, \gamma = \grave{a}\,\grave{\varepsilon}\,\varphi\,(a,\varepsilon.) \,\supset_{\varphi}\,.\,\varphi\,(\beta,\gamma))^{3}$$

XV/11 [xxxvi/11]   RUSSELL to FREGE   29.9.1902

14, Cheyne Walk
Chelsea, S.W. London
29 September 1902

Dear Colleague,

My proposal concerning logical types now seems to me incapable of doing what I had hoped it would do. From Cantor's proposition that any class contains more subclasses than objects we can elicit constantly new contradictions.[1] E.g.: If m is a class of propositions, then '$p \; \varepsilon \; m \cdot \supset_{p} \cdot p$' represents their logical product. This proposition itself can either be a member of class m or not. Let w be the class of all propositions of the above form which are not members of the pertinent class m, i.e.,

$$w = p\,\mathfrak{z}\,(\mathfrak{z}\,m\,\mathfrak{z}\,\{p\,. = : q\,\varepsilon\,m\,.\supset_{q}.q :.\,p \sim \varepsilon\,m\});$$

and let r be the proposition $p \; \varepsilon \; w \cdot \supset_{p} \cdot p$.
We then have $r \; \varepsilon \; w \cdot \equiv \cdot r \sim \varepsilon \; w$.
Here we must consider the content of the propositions, not their meaning; and we must not take equivalent propositions to be simply identical.

Concerning relations, I assume

$$\vdash (R) \sim R\,(R)\,. =_{R} .\,(R)\,T(R): \,\supset : (T) \sim T(T). = (T)\,T(T)$$

---

3 Cf. the last formula in XV/11 below.
XV/11. 1 For Russell's analysis of Cantor's diagonal construction, cf. *Principles*, ch. XLIII, esp. sects 344–50 (pp. 362ff). For the antinomy described here, cf. the more detailed explanations in sect. 500 (pp. 527ff).

This merely shows that the hypothesis is not permitted. To translate the proposition into your conceptual notation, let us assume

$$R = \grave{\alpha}\grave{\varepsilon}\varphi(\alpha, \varepsilon)$$

Then $(R) \sim R\,(R)$ says the following:

$$\longrightarrow\!\!\!\!\top\!\!-\varphi\{\grave{\alpha}\grave{\varepsilon}\,\varphi\,(\alpha,\,\varepsilon),\,\grave{\alpha}\grave{\varepsilon}\,\varphi\,(\alpha,\varepsilon)\}$$

Let us regard φ here as variable, and let us consider the function

$$\top\!\!\!-\!\!\!\!\begin{array}{l}\varphi\{\grave{\alpha}\grave{\varepsilon}\varphi\,(\alpha,\,\varepsilon),\,\beta\}\\[2pt]\beta = \grave{\alpha}\grave{\varepsilon}\,\varphi\,(\alpha,\,\varepsilon)\end{array}$$

or better

$$\top\!\!\!-\!\!\!\!\begin{array}{l}\varphi\,(\beta,\gamma)\\[2pt]\beta = \grave{\alpha}\grave{\varepsilon}\,\varphi\,(\alpha,\varepsilon)\\[2pt]\gamma = \grave{\alpha}\grave{\varepsilon}\,\varphi\,(\alpha,\varepsilon)\end{array}$$

Let us take into consideration the double range of values T of those values of β and γ for which the above function, for any value of φ, is the true; that is,

$$T = \grave{\beta}\grave{\gamma}\;\;\neg\!\!\cup^{\varphi}\!\!\top\!\!\!-\!\!\!\!\begin{array}{l}\varphi(\beta,\,\gamma)\\[2pt]\beta = \grave{\alpha}\grave{\varepsilon}\,\varphi\,(\alpha,\,\varepsilon)\\[2pt]\gamma = \grave{\alpha}\grave{\varepsilon}\,\varphi\,(\alpha,\,\varepsilon)\end{array}$$

i.e.,

$$T = \grave{\beta}\grave{\gamma}\;\;\neg\!\!\cup^{\varphi}\!\!\top\!\!\!-\!\!\!\!\begin{array}{l}\varphi\,(\beta,\,\gamma)\\[2pt]\beta = \grave{\alpha}\grave{\varepsilon}\,\varphi\,(\alpha,\,\varepsilon)\\[2pt]\gamma = \grave{\alpha}\grave{\varepsilon}\,\varphi\,(\alpha,\,\varepsilon)\end{array}$$

and let us ask ourselves whether $-T\cap(T\cap T)$ is the true or the false. (I hope I have written this correctly, for I still have some difficulty writing your conceptual notation.) What this is supposed to bring out is that not only functions but also ranges of values belong to different types. If my theory of types is correct, there must be infinitely many types; but I do not know yet whether there must be more than $\tilde{\omega}$ of them.

Yours sincerely,
Bertrand Russell

XV/12 [xxxvi/12]   FREGE to RUSSELL   20.10.1902

Jena

20 October 1902

Dear Colleague,

Your example of propositions like '$p \, \varepsilon \, m \cdot \supset \cdot p$' prompts me to ask the question: what is a proposition? German logicians understand by it the expression of a thought, a group of audible or visible signs expressing a thought. But you evidently mean the thought itself. This is how mathematicians tend to use the word. I prefer to follow the logicians in their usage. As you know, I distinguish between the sense and the meaning of a sign, and I call the sense of a proposition a thought and its meaning a truth-value. All true propositions have the same meaning: the true; and all false propositions have the same meaning: the false. When I assert that $3 + 5 > 7$, I assert that there is a certain relation between the meanings of the signs '3 + 5' and '7', and these meanings are numbers. But in saying something about the meaning of the sign '3 + 5', I express a sense, or thought. And part of this thought is not the meaning of the sign '3 + 5' but its sense. Likewise, the sense of '3', the sense of '+', and the sense of '5' are parts of the sense of '3 + 5'. The object about which I am saying something – what I mean, what I understand by the sign – is always the meaning of the sign; but in saying something about it I express a thought, and the sense of the sign is part of this thought. Thus what I am talking about when I use a sign is not the sense of the sign. But it can happen that I want to talk about the sense, e.g., about a certain thought. This happens in indirect speech. In the period 'Aristotle believed that the velocity of a falling body was proportional to the time of its fall' what we have in the subordinate clause is indirect speech. What would be the sense of this clause if it was the main clause is now its meaning. I can say: here the subordinate clause is the proper name of a thought, just as 'Aristotle' is the proper name of a philosopher. The subordinate clause does not here express a thought but designates a thought. In my *Conceptual Notation* I did not yet introduce indirect speech because I had as yet no occasion to do so. In direct speech I can always interchange '3 + 5' and '10 − 2', for since both have the same meaning, the meaning of the whole proposition, its truth-value, remains unchanged. Its sense, the thought, is of course changed. That is why I cannot in general interchange '3 + 5' and '10 − 2' in indirect speech, for here the meaning is the thought and this is changed; which will perhaps change the meaning of the whole period, part of which is in indirect speech. Now what does Peano understand by 'proposition': the thought or the truth-value? I believe that he himself does not know and that he sometimes understands the one and sometimes the other by it. If by '2' Peano understands a number, then by '3 > 2' he cannot understand a thought; for '$2^2 - 1^2$' means the same as '3'. Now if the thought was the meaning of '3 > 2', then it would not change if '3' was replaced by '$2^2 - 1^2$'; for there is no change in the meaning. But the thought is indeed changed. Its relation to the proposition '3 > 2' is therefore quite different from the relation of numbers to their signs. In direct speech a proposition is not the sign of a thought but its expression. But the number signs are signs of numbers. It seems to me that on this point Peano is overcome by obscurity and that for this reason he cannot quite explain his sign '⊃'. I find that he gives different explanations and leaves it unclear how they are logically related to one another; which is why one cannot operate safely with the sign '⊃'. When you speak of a class of propositions, you seem to mean a class of thoughts; but in '2 > 1 · ⊃ · 2² > 1', '2 > 1' must be understood to mean the truth-value, and hence the true.

I now believe I understand what you say about relations. I am thinking of coming back to this.

I now avoid the contradiction you have uncovered by replacing my fundamental law (V) by:

$$\vdash (\acute\varepsilon f(\varepsilon) = \acute{a} g(a)) = \left(\begin{array}{l} \smallsmile^{a} \!\!\! \rule{0pt}{0pt} \!\!\! -f(a) = g(a) \\ \qquad a = \acute\varepsilon f(\varepsilon) \\ \qquad a = \acute{a} g(a) \end{array} \right)$$

(Va) can remain unchanged. (Vb) is to be replaced by:

$$\begin{array}{l} -f(a) = g(a) \\ \vdash\ a = \acute\varepsilon f(\varepsilon) \\ \ \ \acute\varepsilon f(\varepsilon) = \acute{a} g(a) \end{array} \qquad \text{or} \qquad \begin{array}{l} -f(a) = g(a) \\ \vdash\ a = \acute{a} g(a) \\ \ \ \acute\varepsilon f(\varepsilon) = \acute{a} g(a). \end{array}$$

Accordingly, one concept can have the same extension as another even though this extension falls under the one and not under the other. It is only necessary that all objects other than the extension itself which fall under the one concept also fall under the other, and conversely.[1] Then:

$$\vdash\ a \frown a$$

Yours sincerely,
G. Frege

XV/13 [xxxvi/13]   RUSSELL to FREGE   12.12.1902

14, Cheyne Walk
Chelsea, S.W.
12 December 1902

Dear Colleague,

When I wrote about the propositions $p \, \varepsilon \, m \cdot \supset \cdot p$, I understood by a *proposition* its sense, not its truth-value. I cannot bring myself to believe that

---

XV/12. 1 The proposed change is carried out in GGA II, p. 262b. There the formulas cited here carry the designations (V'), (V'b), and (V'c).

the true or the false is the meaning of a proposition in the same sense as, e.g., a certain person is the meaning of the name Julius Caesar. But this is an incidental matter. It must be admitted that there are different senses, and it is to be supposed that classes of senses have numbers.[1] Now the sense of '$p \, \varepsilon \, m \cdot \supset \cdot p$' stands in a one-one relation to m; consequently, there is the same number of senses as there is of classes of senses. (For m must be a class of senses; otherwise '$p \, \varepsilon \, m \cdot \supset \cdot p$' has no sense.)

I must give more thought to your solution of the contradiction. There have been many demands on my time recently, and I find it difficult to get on with fundamental questions. But I find it difficult to accept your solution even though it is probably correct. Do you deny, e.g., that all classes form a class? And if this is admitted, then it is possible that $-a \frown a$. Moreover, the class of non-humans is a non-human. Otherwise it must be admitted that not all objects fall either under a or under not-a; namely, if a is a range of values, then a falls neither under a nor under not-a. This contradicts the law of excluded middle, which will be inconvenient to say the least.

I hope that the second volume of your *Arithmetic* will appear soon. Did you see my article in *Revue de mathématiques* 7, 8,[2] as well as an article by Whitehead, *American Journal of Mathematics*, Oct. 1902 (vol. 24)? In the latter, I gave the same definition of a cardinal number that you gave: at the time I did not yet know that you had already made this discovery. The definitions of 1, 2, 3 are false; they should read:

$$1 = u \, \vartheta \left\{ \, \exists u \cap x \, \vartheta \, (u - \iota x \, \varepsilon \, 0) \right\} \text{ Df etc.}^{3}$$

Yours sincerely,
Bertrand Russell

XV/13. 1 Evidently in the sense of 'cardinal number'.

2 B. Russell, 'Sur la logique des relations avec des applications à la théorie des séries', *Rivista di matematica* (= *Revue de mathématiques*) 7 (1900–1), pp. 115–48 (1901), and 'Théorie générale des séries bien ordonnées', ibid. 8 (1902–6), pp. 12–43 (1902).

3 A. N. Whitehead, 'On Cardinal Numbers', loc. cit., pp. 367–94. Section III ('On Finite and Infinite Cardinal Numbers') begins with the words: 'This section III is entirely due to Russell, with the exception of some of the notes' (p. 378). When Russell proposes to correct the definitions of the individual cardinal numbers 1, 2, 3, he is presumably referring to the definitions given in that section. The definition of 1, for example,

$$1 = \text{cls} \cap u \, \vartheta \, (x \varepsilon u . \supset . u - \iota x \, \varepsilon \, 0)$$

needs to be corrected because the existential requirement expressed by '$\exists u$' is missing, and as a consequence, the defining propositional function '$x \, \varepsilon \, u \cdot \supset \cdot u - \iota x \, \varepsilon \, 0$', which is equivalent (in a current notation) to '$\wedge_x \cdot x \varepsilon u \rightarrow u \subset \{x\}$', is satisfied not only by unit classes, as it was intended to be, but also by the null class.

XV/14 [xxxvi/14]   FREGE to RUSSELL   28.12.1902

Jena
28 December 1902

Dear Colleague,

You could not bring yourself to believe that the truth-value is the meaning of a proposition. I do not know whether you read my essay on sense and meaning in vol. 100 of the *Zeitschrift für Philosophie und philosophische Kritik*. The distinction between the sense and the meaning of a sign is important in our case too. It frequently happens that different signs designate the same object but are not necessarily interchangeable because they determine the same object in different ways. It could be said that they lead to it from different directions. The words 'morning star' and 'evening star' designate the same planet, Venus; but to recognize this, a special act of recognition is required; it cannot simply be inferred from the principle of identity. Wherever the coincidence of meaning is not self-evident, we have a difference in sense. Thus the sense of '$2^3 + 1$' is also different from the sense of '3^2' even though the meaning is the same, because a special act of recognition is required in order to see this. Thus the equations '$3^2 = 3^2$' and '$2^3 + 1 = 3^2$' do not have the same cognitive value even though their truth-value is the same. The difference is one of sense: the thoughts expressed are different. If a thought were the meaning of a proposition, then it would not change if one of its parts was replaced by another expression with the same meaning. I now ask: does the whole proposition only have a sense, or does it also have a meaning? What we talk about is the meanings of words. We say something about the meaning of the word 'Sirius' when we say: 'Sirius is bigger than the sun'. This is why in science it is of value to us to know that the words used have a meaning. Of course, in poetry and legend it makes no difference to us. When we merely want to enjoy the poetry we do not care whether, e.g., the name 'Ulysses' has a meaning (or, as it is usually put, whether Ulysses was an historical personage). The question first acquires an interest for us when we take a scientific attitude – the moment we ask, 'Is the story true?', i.e., when we take an interest in the truth-value. In poetry too there are thoughts, but there are only pseudo-assertions. This is also why a poet cannot be accused of lying if he knowingly says something false in his poetry. Now it would be impossible to see why it was of value to us to know whether or not a word had a meaning if the whole proposition did not have a meaning and if this meaning was of no value to us; for whether or not that is so does not affect the thought. Moreover, this meaning will be something which will have value for us precisely when we are interested in whether the words are meaningful, and hence, when we inquire about truth. The meaning of the proposition must be something which does not change when one sign is replaced by another with the same meaning but a different sense. What does not change in the process is the truth-value. If the sign of identity is used between propositions, then the truth-value must be recognized as the meaning of the proposition (indirect speech calls for special consideration). The propositions 'the morning star is a planet' and 'the evening star is a planet' do not have the same sense; but the latter arises from the former if the proper name 'morning star' is replaced by 'evening star', a proper name with the same meaning. It follows from this that the propositions must agree in their meanings: (The morning star is a planet) = (The evening star is a planet) according to the law

$$\vdash \begin{matrix} F(a) = F(b) \\ a = b \end{matrix}$$

This shows that the thought expressed by the proposition cannot be what is recognized as the same, any more than the sense of 'morning star' can be supposed to coincide with the sense of 'evening star' when we write 'morning star = evening star', or than the equation '$2^3 + 1 = 3^2$' can be supposed to mean that the sense of '$2^3 + 1$' coincides with the sense of '3^2'.

Now you use the word 'proposition' in the sense in which I use the word 'thought' or 'sense of a proposition'. A proposition is for me a composite sign which is supposed to express a thought. Your usage would coincide with that of the mathematicians, mine with that of the logicians. I prefer the latter, because otherwise we lack a word for 'proposition' in my sense.

By '$p \, \varepsilon \, m \cdot \supset \cdot p$' you therefore designate a thought – the thought namely that *all thoughts belonging to class m are true*, if I understand your signs correctly. But now a class cannot be a component part of a thought, though the sense of a class name can. You use the proposition as in indirect speech. But in indirect speech (*oratio obliqua*) every word has not its ordinary (direct) meaning but, as I put it, its indirect meaning, which coincides with what is otherwise its sense. Let, e.g., 'M' and 'N' (in direct speech) be names of the same class, so that $M = N$ is the true. But I assume that these names have a different sense in that they determine the class in different ways. Then M cannot be interchanged with N in indirect speech because their meanings there are different. To avoid ambiguity, we ought really to have special signs in indirect speech, though their connection with the corresponding signs in direct speech should be easy to recognize. The thought that *all thoughts belonging to class M are true* is different from the thought that *all thoughts belonging to class N are true*; for someone who did not know that M coincided with N could hold one of these thoughts to be true and the other to be false. Here, in indirect speech, the expression 'the class M' does not therefore mean a class; for then it would mean the same class as 'N'. If we now go on to say, 'The thought that *all thoughts belonging to class M are true* belongs to class M', then M, in its second occurrence, has its direct or ordinary meaning; it here means a class, whereas in the underlined part[1] everything has an indirect meaning. If we replace 'M' in its second

---

XV/14. 1 i.e., in what is printed in italics here.

occurrence by '*N*', the truth-value remains the same. But if we do the same thing in the underlined part, the truth-value may be reversed; for even if the thought that *all thoughts belonging to class M are true* belongs to class *M*, the thought that *all thoughts belonging to class N are true* need not belong to class *M*; for it is a different thought. Since '*M*' has different meanings in its two occurrences in the proposition 'the thought that *all thoughts belonging to class M are true* does not belong to class *M*', there must also be a difference in the meanings of '*M*' in the expression 'the thought that *the thought that ALL THOUGHTS BELONGING TO CLASS M ARE TRUE does not belong to class M*'. It can be said that in the twice-underlined part[2] it has an indirect meaning of the second degree, whereas in the once-underlined part it has an indirect meaning of the first degree. This greatly complicates the matter; and if by '*w*' you now designate the class of all thoughts expressed in the form 'that *the thought that ALL THOUGHTS BELONGING TO CLASS m ARE TRUE does not belong to class m*', and if by '*r*' you designate the thought that *all thoughts belonging to class w are true*, then I do not quite know how you arrive at the equation '$r \varepsilon w \cdot = \cdot r \sim \varepsilon w$', nor what it is supposed to mean: whether a coincidence of thoughts or of truth-values. By what methods of inference do you get your equation?

I once read something of yours in the *Revue de mathématiques*, which showed me that we sometimes followed the same or similar lines of thought. I do not know your article in the *American Journal of Mathematics*.

You will have received the second volume of my *Basic Laws*.

With best wishes for your well-being in the new year,

Yours sincerely,
G. Frege

XV/15 [xxxvi/15]   RUSSELL to FREGE   20.2.1903

14, Cheyne Walk
Chelsea, S.W.
20 February 1903

Dear Colleague,

I cannot remember whether I expressed my thanks to you for your second volume; in case I did not, please excuse my tardiness. Until now I have not been able to read the whole, but I have the intention of familiarizing myself with it shortly. Your criticism of the arithmetical theory of irrational numbers seems to me fully justified; yet I hold a purely arithmetical theory which does not suffer from any logical errors. Let *k* be a class of rational numbers; I then call the class of all rational numbers smaller than at least one member

---

2 i.e. the text printed in italic capitals.

of k the real number determined by k. Some hints of this theory are to be found in Peano.[1]

What you say about my contradiction is of the greatest interest to me. Do you believe that the range of values remains unchanged if some subclass of the class is assigned to it as a new member?[2] Extension seems to fit this view better than intension. But I feel far from clear about this question.

I have read your essay on sense and meaning, but I am still in doubt about your theory of truth-values, if only because it appears paradoxical to

---

XV/15. 1 Russell is here referring to Peano's introduction of 'segments' in his article 'Sui numeri irrazionali', *Rivista di matematica* 6 (1896–1900), pp. 126–40 (1899), esp. p. 133, as well as to Peano's *Formulaire de mathématiques* II (Turin 1899), p. iii. Cf. Russell, *Principles*, ch. XXXIII (pp. 270–5), esp. the note on pp. 274ff, in which Russell criticizes Peano's conception of the relationship between the theory of segments and the theory of irrational numbers. On p. 271, n., Russell traces the idea of using segments for defining real numbers back to hints by G. Cantor, in 'Beiträge zur Begründung der transfiniten Mengenlehre' (first article), *Mathematische Annalen* 46 (1895), pp. 481–512. However, the idea seems to go back to Moritz Pasch: in his *Einleitung in die Differential- und Integralrechnung* (Leipzig 1882), Pasch introduces 'number segments' as groups (= sets) of rational numbers satisfying the following conditions:

'(1) the group [of numbers] does not comprise all numbers;

(2) if number x belongs to the group, all smaller numbers also belong to it;

(3) there is no greatest number in the group' (ibid., p. 203).

Pasch regards the use of these number segments for introducing irrational numbers as a further development of Dedekind's procedure: 'The following introduction of irrational numbers rests on the views developed by R. Dedekind in his work *Stetigkeit und irrationale Zahlen* (Brunswick 1872) [E.T. in *Essays on the Theory of Numbers* (Chicago 1901)]' (ibid., p. 1, n.).

2 This question was presumably prompted by the following sentence in Frege's postscript:

It might at first be feared that according to our stipulations concepts with the same extension would have to be assigned the same number even though there was one object more falling under the one than under the other concept, namely the extension itself, so that in the end we should get only a single finite number (GGA II, p. 264b).

Frege removes this doubt by pointing to the exact formulation of the definition of number. In XV/16 below, Frege replies to Russell's question that a class does *not* in general remain unchanged when *any* subclass is added to it, but that a concept $\Phi(\xi)$ under which its own extension does not fall has the same extension as a concept under which all objects falling under $\Phi(\xi)$ fall and, moreover, the extension $\dot{\varepsilon}(\Phi(\varepsilon))$. Russell seems to proceed as always on the assumption that a class can be a member of itself:

$$ - \dot{\varepsilon}\Phi(\varepsilon) \frown \dot{\varepsilon}\Phi(\varepsilon), $$

and in the present case on the assumption that $\dot{\alpha}(\alpha = \dot{\varepsilon}\,\Phi(\varepsilon))$, which contains $\dot{\varepsilon}\,\Phi(\varepsilon)$ as its one and only element, is a subclass of $\dot{\varepsilon}\,\Phi(\varepsilon)$. Cf. the rest of Frege's discussion in GGA II, pp. 264ff.

me. I believe that a judgement, or even a thought, is something so entirely peculiar that the theory of proper names has no application to it. But there are certain difficulties which I cannot dispose of.

Concerning '$p \, \varepsilon \, m \cdot \supset \cdot p$', I believe that the class m is itself a component part of this thought. If this is impossible, then your criticism is justified; but the impossibility is not evident to me. Let me now put I' for 'is identical with'. We then have

$$w = r \mathfrak{z} (\mathfrak{I} m \mathfrak{z} \{ r \, \mathsf{I}' (p \varepsilon m . \supset_p . p) . r \sim \varepsilon m \}) . \supset ::: (q \varepsilon w . \supset_q . q) \varepsilon w . \supset .$$
$$\mathfrak{I} m \mathfrak{z} ((q \varepsilon w . \supset_q . q) \, \mathsf{I}' (p \varepsilon m . \supset_p . p) . (q \varepsilon w . \supset_q . q) \sim \varepsilon m) \quad (1).$$
$$(q \varepsilon w . \supset_q . q) \, \mathsf{I}' (p \varepsilon m . \supset_p . p) . \supset . w \, \mathsf{I}' m \quad (2)$$
$$(1) . (2) . \supset : (q \varepsilon w . \supset_q . q) \varepsilon w . \supset . (q \varepsilon w . \supset_q . q) \sim \varepsilon w \quad (3)$$

Similarly: $(q \varepsilon w . \supset_q . q) \sim \varepsilon w . \supset . (q \varepsilon w . \supset_q . q) \, \varepsilon w.$

Hence the contradiction.

With best wishes,

Yours sincerely,
Bertrand Russell

XV/16 [xxxvi/16]   FREGE to RUSSELL   21.5.1903

Jena
21 May 1903

Dear Colleague,

Excuse me for answering your letter of 20 February only now; various occupations have kept me from it. In the meantime I have received your *Principles of Mathematics*, vol. I, for which I thank you very much. Although I have not yet got around to taking more than a casual look at it, it has aroused a lively interest in me, and I am anxious to gain a more thorough knowledge of its contents soon. I note with satisfaction that you have devoted a special appendix to my theories. This will contribute greatly to making them more widely known and, I hope, to the advancement of science. The theory of irrational numbers you hint at in your letter seems to me logically unassailable, as long as the word 'class' is acknowledged to have a meaning. But otherwise it will not be possible at all to give arithmetic a logical foundation. When you define an irrational number as a class of rational numbers, it is, of course, something different from what I call an irrational number according to my definition, although there is naturally a connection. It seems to me that you need a double transition: (1) from numbers to rational numbers, and (2) from rational to real numbers in general. I want to go at once from numbers to real numbers as relations of magnitudes.[1]

XV/16. 1 Cf. note 1 to XV/15 above.

I do not believe that a class remains in general unchanged when a particular subclass is added to it. All I mean is that two concepts have the same extension (the same class) when the only difference between them is that this class falls under the first concept but not under the second. But I do not know if I properly understood your question.[2]

If I understand you aright, by 'p' in '$p \, \varepsilon \, m \supset p$' you indicate a thought, and by 'm' a class of thoughts; i.e., the letter 'p' stands for the proper name of a thought, and so does the combination of signs '$p \, \varepsilon \, m$'. The thought here is therefore not the sense but the meaning of the sign. If '$p \, \varepsilon \, m$' expresses a thought, then the sense of 'm' will be a component part of this thought. But now m is supposed to be a class; the class is the meaning, not the sense of 'm', and the thought here is not supposed to be expressed but to be designated. Now under these circumstances is the class a component part of the thought? If a proposition does not express a thought but designates one, then all parts of the proposition are to be taken with their indirect meanings. The indirect meaning of a sign is its sense in its ordinary use. Now a class cannot be the sense of a sign, but only its meaning, as Sirius can only be the meaning of a sign, but not its sense. Hence a class cannot be the indirect meaning of a sign, any more than Sirius can. Accordingly, if '$p \, \varepsilon \, m$' is supposed to designate a thought, m cannot be a class. Can any class whatever be a component part of a thought? No more than the planet Jupiter can. A class (or the corresponding concept) can be defined in different ways, and to a different definition corresponds a different sense of the class name. Now the thought that an object belongs to a class as defined in one way is different from the thought that an object belongs to a class as defined in another way. Consequently, a class cannot itself be part of the thought that an object belongs to it (for the class is the same in both cases); but only the sense of the class name can be part of this thought. If the class was part of the thought that an object p belonged to it, then the change in the sense of the class name would not affect the thought, provided that the class itself remained unchanged.

We must now come to an understanding about certain fundamental questions: is a thought the sense of a proposition, or is it its meaning? In other words: does a proposition express a thought, or does it designate it? I shall disregard indirect speech here, for since it occurs in a subordinate clause, it is only a dependent part of the propositional whole. If one does not recognize that the meaning of a proposition is its truth-value, the question arises, Does a proposition have no meaning at all, and is the thought its sense? Or does a proposition have a meaning, and is this a thought? If the latter were the case, then the propositions '$2^3 > 7$' and '$3^2 - 1 > 7$' would have to designate the same thought; for '2^3' has the same designation as '$3^2 - 1$'. Now the thoughts contained in those propositions are evidently

---

2 Cf. note 2 to XV/15 above.

different; for after having recognized the first as true, we still need a special act to recognize the second as true. If we had the same thought, there would be no need for two acts of recognition but only for a single one. We are thus compelled to regard a thought as the sense of a proposition. Now does the proposition only have a sense, or does it also have a meaning? In the former case, the meaning of any of its parts would make no difference to the proposition; for all that needs to be considered in considering the sense of a proposition is the sense and not the meaning of its parts. Conversely, if the meaning of a part is essential to a proposition, we must consider something about the proposition other than its sense; and what can this be other than its truth-value?

Does the sign '$\supset$' in '$3 > 2 \cdot \supset \cdot 3^2$' designate a relation between thoughts? Then '$3 > 2$' would have to designate a thought, which is impossible, in so far as '2' designates a number; for while the sense of a number sign can be part of a thought, a number itself cannot. We can only be dealing with a relation between the meanings of '$3 > 2$' and '$3^2 > 2$', which must therefore exist and which cannot be thoughts. What is the meaning of '$\supset$' in Peano? We are given different explanations. What is a *proposition* in Peano? If we give clear answers to these questions that fit all cases, we are driven inevitably towards my view. If I remember aright, Peano has a proposition '$a, b \ \varepsilon \ p \cdot a \supset b \cdot b \supset a : \supset : a = b$' or something like it corresponding to my proposition (IVa), from which it follows that all propositions that express a true thought mean the same, and likewise all propositions that express a false thought. We have, e.g., $3 > 2 \cdot \supset \cdot 2^2 = 4$ and $2^2 = 4 \cdot \supset \cdot 3 > 2$; consequently: $3 > 2 \cdot = \cdot 2^2 = 4$.

Yours sincerely,
G. Frege

XV/17 [xxxvi/17]   RUSSELL to FREGE   24.5.1903

Churt, Farnham
24 May 1903

Dear Colleague,

I received your letter this morning, and I am replying to it at once, for I believe I have discovered that classes are entirely superfluous. Your designation $\dot{\varepsilon}\varphi(\varepsilon)$ can be used for φ itself, and $x \frown \dot{\varepsilon}\varphi(\varepsilon)$ for $\varphi(x)$. (I write ε instead of $\frown$, like Peano.) I put '$p \supset q$' meaning 'p is not true, or q is true', which is the same as:

$$\mathord{\vdash}^q_p$$

Then $p \equiv q \cdot = \cdot p \supset q \cdot q \supset p$ Df ('$=$ ... Df' is a single symbol).

$\varphi \supset \psi \cdot = \cdot \varphi x \supset_x \psi x$ Df

$\varphi \; ||| \; \psi \cdot = \cdot \varphi x \equiv_x \psi x$ Df

$x \, l' y \cdot = \cdot \varphi \supset_\varphi \varphi y$ Df (l' = identity)

'Indiv (x)' means 'x is an object, i.e., not a function'.

$$\sim p \cdot = : p \cdot \supset \cdot (r) \cdot r \, \text{Df}$$

(By '$(x) \cdot \varphi(x)$' I designate the thought that $\varphi(x)$ is true for all values of x, and $\sim p$ designates 'p is not true'.)

$$u = v \cdot = : \cdot \text{Indiv}(u) \cdot \supset \cdot u \, l' \, v : \sim \text{Indiv}(u) \cdot \supset.$$

$$\sim \text{Indiv}(v) \; \cdot u \; ||| \; v \; \text{Df}$$

Equality so defined can be used in enumeration just like identity. If we take, e.g., $(\text{Nc} \to 1) \, (f)$ (which you designate by '$1f$'), we have

$(\text{Nc} \to 1)(f). = \; : f(x,y) . f(x,z) \; . \supset_{x, \, y, \, z} . y = z$ Df

$(1 \to \text{Nc})(f) = \; : f(x,z) . f(y,z) . \supset_{x, \, y, \, z} . x = y$ Df

$(1 \to 1)(f) . = \; . \; (\text{Nc} \to 1)(f) . (1 \to \text{Nc})(f)$ Df

$(\exists x) . \varphi(x) . = \; . \sim \{ (x) . \sim \varphi(x) \}$ Df

$\varphi \; \text{sim} \; \psi . = \; . \; (\exists f). \; \{ (1 \to 1)f : \varphi x \; . \supset. \; (\exists y) . \{ f(x,y) . \psi y \} : \psi y . \supset . \; (\exists x) . \{ f(x,y) . \varphi x \} \}$ Df

$\text{Nc}(\varphi) = \psi' \; (\varphi \; \text{sim} \; \psi)$ Df

Here $\text{Nc}(\varphi)$ means the cardinal number of φ. We have

$$\vdash \, : \varphi \; \text{sim} \; \psi \; . \equiv \; . \; \text{Nc}(\varphi) = \text{Nc} \; (\psi)$$

In this way we can do arithmetic without classes. And this seems to me to avoid the contradiction.

I still do not quite share your opinion about sense and meaning. I should like to say the following about them. In all cases, both imagination and judgement have an object: what I call a 'proposition' can be the object of judgement, and it can be the object of imagination. There are therefore two ways in which we can think of an object, in case this object is a complex: we can imagine it, or we can judge it; yet the object is the same in both cases (e.g., when we say 'the cold wind' and when we say 'The wind is cold'). To me, the judgement stroke therefore means a different way of being directed towards an object. Complexes are true or false: in judging, we aim at a true complex; but we may, of course, miss our aim. But truth is not a component part of the true, as *green* is a component part of a green tree. For me there is nothing identical about two propositions that are both true or both false; I write $p \equiv q \cdot = \cdot p \supset q \cdot q \supset p$ Df. This relation exists accordingly between any two objects which are not meanings of propositions. Concerning '$p \, \varepsilon \, m \cdot \supset_p \cdot p$', I now write '$\varphi(p) \supset_p p$' instead, i.e., 'The property φ does not belong to any objects which are not true'. It seems to me that there is no difficulty in this, provided one does not require '$p \equiv q$' to express an identity. The function which gave rise to difficulties for me was $\sim \varphi \, \{ \varphi(p) \supset_p p \}$. But now these difficulties have been overcome by means of the theorem in your appendix, according to which[1]

$$\vdash : \exists \, (\varphi, \psi) \cdot ((\varphi(p) \supset_p p) \, \mathsf{I}' \, (\psi \, (p) \supset_p p) \cdot \sim \varphi \, \{ \, \varphi(p) \supset_p p \, \} \cdot \psi \\ \{ \psi \, (p) \supset_p p \})$$

In the first two parts of my book there are many things which I did not discuss thoroughly, and many opinions which no longer seem to me correct. But in the later parts most things seem to me correct, provided classes are everywhere replaced by functions. In the second volume I hope to work out everything in symbols.[2]

Yours sincerely,
Bertrand Russell

XV/18 [xxxvi/18]   FREGE TO RUSSELL   13.11.1904[1]

Jena
13 November 1904

Dear Colleague,

Excuse me if, due to various distractions, I have only now got around to answering your letter of 24 May of last year.

I cannot regard your attempt to make classes entirely dispensable as successful, the reason being that you use function letters in isolation. I have

---

XV/17. 1 Russell is evidently referring to GGA II, p. 261. There Frege first derives a proposition ω, and by replacing 'a' by a combination of signs '$M_\beta \, (\Psi(\beta))$' (formed by means of an abbrevation '$\Psi(\beta)$' which is expressly introduced for this purpose), he then gets a proposition (not expressed in his conceptual notation) about the existence of two concepts 'such that, taken as arguments of the second-level function, they yield the same value which now falls under the second of these concepts but not under the first' (ibid., p. 261b). In order to avoid confusion with the 'φ' used by Russell at this point, let us replace the standard notation used in GGA I, p. 42, i.e., '$M_\beta \, (\varphi(\beta))$', by '$M_\beta \, (\chi(\beta))$'. For $M_\beta \, (\chi(\beta))$ Russell puts the function $\chi(p) \supset_p p$; his 'φ' then corresponds to Frege's 'Ψ', and his 'ψ' to the function whose existence is asserted in the conclusion Frege draws from proposition ω and which is indicated by '$\mathfrak{g}$'.

2 The second volume of *The Principles of Mathematics*, which is being announced here and to which the first volume was to be only a 'commentary or introduction' according to Russell's preface, never appeared as such. Conceived from the outset as a joint work by Alfred North Whitehead and Bertrand Russell, it was begun in the year 1900 and published as *Principia Mathematica* I–III (1910–13). Cf. B. Russell, *The Principles of Mathematics* (Cambridge 1903), p. vi, or 2nd ed. (London 1937), p. xvi, and A. N. Whitehead and B. Russell, *Principia Mathematica* I (Cambridge 1910), p. v.

XV/18. 1 There exists a 17-page draft of this letter in Frege's handwriting. With the exception of two places considered in notes 4 and 5, it does not differ essentially from the final text of the letter.

expressed my view on this in n. 2 on p. 2 of vol. II of my *Basic Laws*. To use a function sign in isolation is to contradict the nature of a function, which consists in its unsaturatedness. For this is how a function differs from an object. This is also why function names must differ essentially from proper names, the difference being that they carry with them at least one empty place – an argument place. And these argument places must always be preserved in a function name and be recognizable as such; otherwise the function name becomes a meaningless proper name. The same must hold for function letters, at least wherever they are to be replaceable by function names. I therefore see to it that Roman function letters are always followed by parentheses which contain the argument place in the space between them. In my usage, only the German function letters can occur in isolation, though only where they occur above the concavity, for in this occurrence they can never be replaced by function names. But this is also the only case where there is no possibility of replacement. With your designations '$\varphi \supset \psi$' and '$\varphi \mathbin{\|\|} \psi$', we get at once into a difficulty if we make the transition to particular functions. According to the notation you use in your letter, we have, e.g.,

$$x^2 = 1 \cdot \underset{x}{\supset} \cdot x\,(x-1)\,(x+1) = 0,$$

analogous to your

$$\varphi\,x \underset{x}{\supset} \psi\,x;$$

but we have nothing analogous to your '$\varphi \subset \psi$'. To get this, we would first have to transform all function names* in such a way that there was only one argument place and that it was on the right-hand side. Thus we would have to transform, e.g., '$x^2 = 1$' into '$\dot{\varepsilon}(\varepsilon^2 = 1)x$', and '$x(x-1)\,(x+1) = 0$' into '$\dot{\varepsilon}\,(\,\varepsilon(\varepsilon-1)\,(\varepsilon+1) = 0)x$', so that we could write

$$\dot{\varepsilon}(\varepsilon^2 = 1)x \underset{x}{\supset} \dot{\varepsilon}(\varepsilon(\varepsilon-1)\,(\varepsilon+1) = 0)x$$

and for this, according to your definition,

$$\dot{\varepsilon}(\varepsilon^2 = 1) \subset \dot{\varepsilon}(\varepsilon(\varepsilon-1)\,(\varepsilon+1) = 0).$$

But this notation[2] would lead to the same difficulties as my value-range notation and in addition to a new one. For a range of values is supposedly an object and its name a proper name; but '$\dot{\varepsilon}(\varepsilon^2 = 1)$' would supposedly be a function name which would require completion by a sign following it. '$\dot{\varepsilon}(\varepsilon^2 = 1)1$' would have the same meaning as '$1^2 = 1$', and accordingly, '$\dot{\varepsilon}(\varepsilon^2 = 1) \subset$' would have to have the same meaning as '$\subset^2 = 1$', which, however, would be meaningless. '$\dot{\varepsilon}(\varepsilon^2 = 1)$' would be defined only in connection with an

---

* Names of first-level functions with one argument.

---

2 To distinguish '$\dot{\varepsilon}\,\Phi(\varepsilon)x$' from his value-range notation '$\acute{\varepsilon}\,\Phi(\varepsilon)$', Frege here uses the *spiritus asper* instead of the *spiritus lenis* over the initial 'ε'.

argument sign following it, and it would nevertheless be used without one; it would be defined as a function sign and used as a proper name, which will not do.

In your designations '$\varphi \parallel \psi$' and '$\varphi = \psi$', 'φ' and 'ψ' are no longer used as function letters but as object letters. Given the notation assumed above, we would have '$\dot{\varepsilon}(\varepsilon^2 = 1) = \dot{\varepsilon}\{(\varepsilon - 1)(\varepsilon + 1) = 0\}$', which does not differ essentially from my '$\dot{\varepsilon}(\varepsilon^2 = 1) = \dot{\varepsilon}\{(\varepsilon - 1)(\varepsilon + 1) = 0\}$'.[3] In this notation we would also have

$$\dot{\varepsilon}\varphi(\varepsilon)x \cdot \supset \cdot \varphi x \quad \text{and} \quad \varphi x \cdot \supset \cdot \dot{\varepsilon}\varphi(\varepsilon)x$$

and further

$$\dot{\varepsilon}(\sim \cdot \varepsilon\varepsilon)\Psi : \supset : \sim \cdot \Psi\Psi \quad \text{and} \quad \sim \Psi\Psi \cdot \supset \cdot \dot{\varepsilon}(\sim \cdot \varepsilon\varepsilon)\Psi$$

and finally

$$\dot{\varepsilon}(\sim \cdot \varepsilon\varepsilon) \, \dot{\varepsilon} \, (\sim \cdot \varepsilon\varepsilon) : \supset : \sim \dot{\varepsilon}(\sim \cdot \varepsilon\varepsilon) \, \dot{\varepsilon} \, (\sim \cdot \varepsilon\varepsilon)$$

and

$$\sim \dot{\varepsilon}(\sim \cdot \varepsilon\varepsilon) \, \dot{\varepsilon} \, (\sim \cdot \varepsilon\varepsilon) : \supset : \dot{\varepsilon} \, (\sim \cdot \varepsilon\varepsilon) \, \dot{\varepsilon} \, (\sim \cdot \varepsilon\varepsilon)$$

and hence the contradiction.[4]

---

3 In the draft of the letter (cf. note 1), Frege first wrote '$\dot{\varepsilon}(\varepsilon(\varepsilon - 1)(\varepsilon + 1) = 0$', as above, in both cases; but the second occurrence of 'ε' has been crossed out in each case.

4 This line of reasoning is intended as a *reductio ad absurdum* of Russell's proposal, as reformulated by Frege, to regard the names of functions as interchangeable, wherever they occur, with the names of their ranges of values. The latter are again written as '$\dot{\varepsilon}\varphi(\varepsilon)$', i.e., with a *spiritus asper*, to distinguish them from Frege's value-range notation (cf. note 2). In detail, Frege's argument proceeds as follows: If function and value-range names are interchangeable, then so are '$\dot{\varepsilon}\varphi(\varepsilon)$' and '$\varphi$' before an object variable 'x'. Hence the implications

$$\dot{\varepsilon}\varphi(\varepsilon)x \cdot \supset \cdot \varphi x \quad \text{and} \quad \varphi x \cdot \supset \cdot \dot{\varepsilon}\varphi(\varepsilon)x$$

hold. Moreover, there are no longer any admissibility restrictions based on the distinction between function and object variables for putting expressions in argument places. Hence in the two formulas just obtained, the variable 'φ' can now both be replaced by the composite function name '$\sim \Psi$' (replacement 1) and by the expression '$\sim \dot{\varepsilon}\Psi(\varepsilon)$' (replacement 2). The latter arises from '$\sim \Psi$' when 'Ψ' is exchanged for '$\dot{\varepsilon}\Psi(\varepsilon)$', which is admissible according to Russell's proposal. Finally, according to the same proposal, the object variable 'x' can now also be replaced by the function name 'Ψ' (replacement 3). Frege carries out these three replacements in the first step in his reasoning, and in the process '$\dot{\varepsilon}\varphi(\varepsilon)$' is transformed into '$\dot{\varepsilon}(\sim \cdot \dot{\varepsilon}\Psi(\varepsilon))$'. Since '$\Psi$' in the (pseudo-) value-range name '$\dot{\varepsilon}\Psi(\varepsilon)$' is no longer supposed to be replaceable, Frege abbreviates the expression '$\dot{\varepsilon}(\sim \cdot \dot{\varepsilon}\Psi(\varepsilon))$' to '$\dot{\varepsilon} (\sim \cdot \varepsilon\varepsilon)$' by omitting the letter 'Ψ'. The first step thus yields the two formulas

$$\dot{\varepsilon}(\sim \cdot \varepsilon\varepsilon) \, \Psi \cdot \supset \cdot \sim \cdot \Psi\Psi \quad \text{and} \quad \sim \cdot \Psi\Psi \cdot \supset \cdot \dot{\varepsilon}(\sim \cdot \varepsilon\varepsilon)\Psi.$$

In the second step. Frege now replaces the letter 'Ψ' by '$\dot{\varepsilon}(\sim \cdot \varepsilon\varepsilon)$', which gives rise to the two formulas at the end. These are of the form $A \supset \sim A$ and $\sim A \supset A$ respectively and hence of the form of Russell's antinomy. This refutation of Russell's proposal should be compared with the formally similar derivation of Russell's antinomy in the postscript to GGA II, p. 257, left column.

You will gather from this that I also cannot take kindly to your designation '(Nc → 1) (f)'.

It is not easy to come to an understanding about what you say about imagination and judgement; it all depends on the sense in which these words are used. It seems to me that what you have in view is the difference between grasping a thought and recognizing a thought as true. The latter is what I call judging. I use 'imagine' and 'imagination' in the psychological sense, not in the logical one, so that these words have no place in mathematics. I agree with you that 'true' is not a predicate like 'green'. For at bottom, the proposition 'It is true that $2 + 3 = 5$' says no more than the proposition '$2 + 3 = 5$'. Truth is not a component part of a thought, just as Mont Blanc with its snowfields is not itself a component part of the thought that Mont Blanc is more than 4000 metres high. But I see no connection between this and what you go on to say: 'For me there is nothing identical about two propositions that are both true or both false'. The sense of the word 'moon' is a component part of the thought that the moon is smaller than the earth. The moon itself (i.e., the meaning of the word 'moon') is not part of the sense of the word 'moon'; for then it would also be a component part of that thought. We can nevertheless say: 'The moon is identical with the heavenly body closest to the earth'. What is identical, however, is not a component part but the meaning of the expressions 'the moon' and 'the heavenly body closest to the earth'. We can say that $3 + 4$ is identical with $8 - 1$; i.e., that the meaning of '$3 + 4$' coincides with the meaning of '$8 - 1$'. But this meaning, namely the number 7, is not a component part of the sense of '$3 + 4$'. The identity is not an identity of sense, nor of part of the sense, but of meaning. You speak of the meaning of a proposition; but what do you understand by it?[5] At another place in your letter you write: 'By "$(x) \cdot \varphi(x)$" I designate the thought that $\varphi(x)$ is true for all values of x'. I would have

---

5 In the draft of the letter (cf. note 1) there follows at this point:
A thought cannot be the meaning of a proposition; for the propositions

$$2^3 - 1 > 2$$
$$3 + 4 > 2$$
$$4^2 - 3^2 > 2$$

have the same meaning because '$2^3 - 1$' has the same meaning as '$3 + 4$' and as '$4^2 - 3^2$'. Now since the propositions agree perfectly in the meanings of their component parts, they must also have the same meaning. [This sentence was subsequently crossed out.] But the thoughts contained in these propositions are different; consequently, they cannot be the meanings. Either we must deny all meaning ['all' crossed out] to these propositions, or we must recognize the truth-value as their meaning. [This sentence was subsequently crossed out.] Do the propositions perhaps have no meaning at all? Then it would have to be immaterial whether or not the constituent parts of the propositions had meanings. But it does matter a great deal. Thus if a proposition has a meaning and if this is not the thought expressed in it, it only remains for us to acknowledge the truth-value as the meaning of the proposition.
The whole paragraph was crossed out by Frege, in some places more than once.

said 'express' instead of 'designate' here. According to my way of speaking, a thought can be designated and it can be expressed. The former happens in indirect speech. 'Copernicus thought that the planetary orbits are circular' is an example of this. The subordinate clause introduced by 'that' designates a thought, while the whole proposition (main clause and subordinate clause) expresses a thought. Copernicus himself was able to express the thought that the planetary orbits are circular. In our whole proposition, the proper name 'Copernicus' designates a man, just as the subordinate clause 'that the planetary orbits are circular' designates a thought; and what is said is that there is a relation between this man and that thought, namely that the man took the thought to be true. Here the man and the thought occupy, so to speak, the same stage. On the other hand, the man and the thought of the whole proposition 'Copernicus thought that the planetary orbits are circular' do not occupy the same stage. If it is said that the name 'Copernicus' here designates a man, then it cannot be said that the whole proposition designates a thought; for the connection between the man and the name is quite different from that between the whole proposition and the thought. The man is designated, the thought is expressed. Moreover, the man is not placed in relation to the thought. Compare this with the following example: '7 − 1' designates a number, just as '7' and '1' designate numbers. These numbers occupy, so to speak, the same stage. The kind of connection between the sign '7 − 1' and the number $7 - 1$ or 6 is the same as that between the sign '7' and the number 7. Now instead of the sign '7' we can also take the sign '4 + 3', and '4 + 3 − 1' now designates the same number as '7 − 1' because '4 + 3' designates the same number as '7'. We can regard $7 - 1$ as a value of the function $\xi - 1$ for the argument 7. And it makes no difference to the value which of the signs '7', '4 + 3', '$4^2 - 3^2$' we use, all of which have the same meaning. In this way, we cannot regard the thought that 7 is greater than 6 as a value of the function $\xi > 6$ for the argument 7; for we get another thought if we substitute '4 + 3' for '7', and yet another if we substitute '$4^2 - 3^2$' for it. We thus find that the thought depends on something other than what is designated by the sign; for this is the same for '7' and for '4 + 3'. A sign must therefore be connected with something other than its meaning, something that can be different for signs with the same designation. Signs do not just designate something; they also express something. This is the sense. Indeed, the two propositions '7 = 7' and

$$\frac{5^2 \cdot 211 - 4}{753} = 7$$

do not have the same cognitive value for us, even though the sign

$$\frac{5^2 \cdot 211 - 4}{753}$$

has the same designation as '7'. The cognitive value therefore does not depend only on the meaning; the sense is just as essential. Without the latter we should have no knowledge at all. When I say '$7 - 1 = 6$', the number 7 does not occupy the same stage as the sense of '$7 - 1$', any more than it occupies the same stage as the thought that $7 - 1 = 6$. On the other hand, the sense of the sign '7' occupies the same stage as this thought; it can be said to be part of this thought, as well as part of the sense of '$7 - 1$'. We must therefore conceive of this thought as the sense of this proposition and say accordingly: the proposition expresses the thought. Now, can we not be satisfied with the sense of the proposition and do without a meaning? For it does sometimes happen that a sign has a sense but no meaning, namely in legend and poetry. Thus the sense is independent of whether there is a meaning. Accordingly, if all that matters to us is the sense of the proposition, the thought, then all we need to worry about is the sense of the signs that constitute the proposition; whether or not they also have a meaning does not affect the thought. And this is indeed the case in legend and poetry. Conversely, if it is not immaterial to us whether the signs that constitute the proposition are meaningful, then it is not just the thought which matters to us, but also the meaning of the proposition. And this is the case when and only when we are inquiring into its truth. Then and only then does the meaning of the proposition enter into our considerations; it must therefore be most intimately connected with its truth. Indirect speech must here be disregarded; for we have seen that, in it, the thought is designated, not expressed. Disregarding it, we can therefore say that any true proposition can be replaced by any true proposition without detriment to its truth, and likewise any false proposition by any false proposition. And this is to say that all true propositions mean or designate the same thing, and likewise all false propositions; and this agrees with your definition:

$$x \; \mathsf{I'} \; y = \varphi x \underset{\varphi}{\supset} \varphi y \; \mathrm{Df}$$

I should like to add some remarks about this. You have the sign of identity '$\mathsf{I'}$' and the equals sign, which you explain as follows:

$$u = v \cdot = \therefore \mathrm{Indiv}(u) \supset u \; \mathsf{I'} \; v : \sim \mathrm{Indiv}(u) \cdot \supset \cdot \sim \mathrm{Indiv}(v) \cdot u \; \mathsf{|||} \; v \; \mathrm{Def.}$$

But here you presuppose that it is known in advance, not only in every definition, but also in the explanation of identity:

$$x \; \mathsf{I'} \; y \cdot = \cdot \varphi x \underset{\varphi}{\supset} \varphi y \; \mathrm{Df}$$

I do not find this permissible. But this only in passing. In the former explanation you distinguish the cases $\mathrm{Indiv}(u)$ and $\sim \mathrm{Indiv}(u)$. Now in your explanation of identity, we have, it seems, the case $\mathrm{Indiv}(x \; \mathsf{I'} \; y)$. If so, you will have to conceive the meaning of any proposition as an object, so that the equals sign, when it occurs between propositions, can be replaced by '$\mathsf{I'}$'. Now the question arises: When does a proposition mean the same object as another proposition? At any rate, '$4^2 - 3^2 = 7$' must mean the same object as '$7 = 7$' because '$4^2 - 3^2$' means the same as '7'. Now what is this object if it is not the truth-value? Incidentally, I do not quite understand the distinction between the equals sign and ' I '. If the equals sign occurs in a definition between the group of signs to be explained and the explicans, it must always be taken as the sign of identity; for it is used to stipulate that that group of signs is to have the same meaning as this one; and this would be the case even if it occurred between isolated function signs, in so far as this isolation was permissible at all. But if it should not be permissible to form a definition in which the equals sign occurred between isolated function signs, then we should, it seems to me, lose the advantage which makes such isolation appear desirable.

If we admit isolated function signs at all, we must also admit them on both sides of the sign of identity. I was interested in your definition '$\sim p \cdot = : p \cdot \supset \cdot (r) \cdot r$'. In my signs it would look like this:

$$\Vdash (\top p) = \quad\vcenter{\hbox{$\overset{a}{\underset{p}{\rule{0pt}{1.2em}}}$}}\, a$$

This would save a primitive sign; but we would probably need some new primitive laws, e.g.:

or $p \supset (r) \cdot r : \supset (r) \cdot r \therefore \supset p$

This is really too complicated for a primitive law; and I do not know whether it can be reduced to something simpler.

Yours sincerely,
G. Frege

XV/19 [xxxvi/19]   RUSSELL to FREGE   12.12.1904

Ivy Lodge
Tilford, Farnham
12 December 1904

Dear Colleague,

Many thanks for your very important letter. I have known already for about a year that my attempt to make classes entirely dispensable was a failure, for essentially the same reasons as you give. But it is not yet clear to me that it is never permissible to use a function letter in isolation. On this point I do not quite share your view, for reasons you will find in my book, sects 480ff, especially sect. 483. My present belief about it is roughly as follows. In the case of a particular function, e.g., $(\xi - 1) \cdot (\xi + 1)$, what arises through the mere omission of ξ can certainly not be regarded as an object. But I believe that if we use the notation φx, the letter φ must designate something that remains the same when y is substituted for x. This something is, I believe, precisely what is designated by $\varphi\xi$. That is, $(\xi-1)\cdot(\xi+1)$ would be a particular value of x. For otherwise it is difficult to find out what is really the same in φx and φy. But I have as yet no definite opinion about this.

I believe that the contradiction does not arise from the nature of a class, but from the fact that certain expressions of the form

$$(\varphi).F(x,\varphi x,\varphi\xi)$$

(where φ should be a German letter) do not represent functions of x. That is, we have

$$\vdash\,::(\exists\,F)\,::\,\sim{}'(\exists f)\,:.\,(x)\,:fx\,.\,\equiv\,.\,(\varphi)\,.\,F\,(x,\,\varphi x,\,\varphi\xi).$$

This is easy to prove in the case of

$$x = \xi'(\varphi\xi)\cdot\underset{\varphi}{\supset}\cdot\sim\varphi x.$$

For this proposition denies $f\{\xi'(f\xi)\}$ for any f; hence no $f\eta$ is always asserted. If we admit functions as objects, then

$$x = \varphi\xi\cdot\underset{\varphi}{\supset}\cdot\sim{}'\varphi x$$

is an even simpler example of this.[1]

XV/19. 1 Russell's line of thought at this point does not become perfectly perspicuous even if the expression that a function is 'always asserted' is interpreted to mean that a propositional function is satisfied by all admissible arguments, an interpretation supported by sect. 103 of the *Principles*. Although the context suggests such a derivation, it does not seem possible to derive Russell's antinomy by means of that expression. For following Russell's proposal, let us take $(\varphi)\cdot F(x,\varphi x,\varphi\xi)$ and choose

$$x = \overset{2}{\xi}(\varphi\xi)\cdot\underset{\varphi}{\supset}\cdot\sim\varphi x,$$

which is an abbreviation of $(\varphi):x = \overset{2}{\xi}(\varphi\xi)\cdot\supset\cdot\sim\varphi x$ (and where '$\overset{2}{\xi}(\varphi\xi)$' corresponds to Frege's value-range notation). If this expression is admissible at all, then it is, as indicated by the universal quantifier, a value of the propositional function fx. By substituting '$\overset{2}{\xi}(f\xi)$' for 'x', we thus get the (true or false) proposition

$$(\varphi):\overset{2}{\xi}(f\xi) = \overset{2}{\xi}(\varphi\xi)\cdot\supset\cdot\sim\varphi(\overset{2}{\xi}(f\xi)).$$

If this proposition was true, then it would have to be true of f, by instantiation, that

$$\overset{2}{\xi}(f\xi) = \overset{2}{\xi}(f\xi)\cdot\supset\cdot\sim f(\overset{2}{\xi}(f\xi)),$$

If I am right in this, then we must come to an understanding about the nature of a function. A proposition of the form

$$x = f(\varphi\xi) \cdot \underset{\varphi}{\supset} \cdot \varphi x$$

asserts those functions for which $x = f(\varphi\xi)$ holds for any x. These functions are usually different for different values of x; this is why it is not self-evident that there is a function which is always asserted. If we accept this as the right way of disposing of the contradiction, then we have the primitive law

$$\vdash :. (x) \cdot fx \equiv gx . \equiv . \overset{2}{\xi} (f\xi) = \overset{2}{\xi} (g\xi),$$

but we need several very complicated primitive laws concerning those cases where we have a functional expression of the form

$$(\varphi) \cdot F(x, \varphi x, \varphi\xi).$$

But it is almost certain that in order to avoid the contradiction we must assume primitive laws which are far from self-evident.

For negation I use as a primitive law:

$$\vdash :. p \supset q . \supset . p : \supset . p$$

which is hardly self-evident.[2] But it is of the same form as the other primitive laws of deduction. These are according to me:

---

and therefore, since $\overset{2}{\xi}(f\xi) \overset{2}{=} \overset{2}{\xi}(f\xi)$ is true, that $\sim f(\overset{2}{\xi}(f\xi))$. But this would contradict the proposition $f(\overset{2}{\xi}(f\xi))$ which is also true on our assumption. Hence $f(\overset{2}{\xi}(f\xi))$ must be false, and so must

$$(\varphi) : \overset{2}{\xi}(f\xi) = \overset{2}{\xi}(\varphi\xi) \cdot \supset \cdot \sim \varphi(\overset{2}{\xi}(f\xi)).$$

However, no contradiction follows from this; it only follows that the given expression is a value of a propositional function which is not satisfied by its own range of values. Russell regards the expression '$x = \varphi\xi \cdot \supset \cdot \sim '\varphi x$' as 'an even simpler example' only because it employs the notation '$\varphi\xi$' (which Frege rejects) instead of the value-range notation '$\overset{2}{\xi}(\varphi\xi)$; hence the very same considerations apply to this expression.

2 This law is introduced by Russell in *Principles*, sect. 18 (pp. 16–17), as axiom 10 under the name of 'principle of reduction'. It is nowadays called 'Peirce's Law' after Charles S. Peirce; cf. his 'On the Algebra of Logic: A Contribution to the Philosophy of Notation', *American Journal of Mathematics* 7 (1885), pp. 180–202. On the intended connection with negation, cf. B. Russell, 'The Theory of Implication', *American Journal of Mathematics* 28 (1906), pp. 159–202, esp. the note on pp. 200ff. Negation is there defined by $\neg p \leftrightarrows p \rightarrow \bigwedge_q q$; but since the statement $\bigwedge_q q$ must be expressly stipulated to be false, one could choose a statement constant $\curlywedge$ ('false') instead of that universal statement and define negation accordingly by $\neg p \leftrightarrows p \rightarrow \curlywedge$. On further pertinent properties of Peirce's Law, cf. the article 'Aussage, Peircesche', in Joachim Ritter (ed.), *Historisches Wörterbuch der Philosophie* I (Basel/Stuttgart 1971), col. 672, and the literature cited in the article.

$$\vdash . p \supset p$$
$$\vdash : p . \supset . q \supset p$$
$$\vdash :. p . \supset . q \supset r : \supset : q . \supset . p \supset r$$
$$\vdash :. p \supset q . \supset : q \supset r . \supset . p \supset r .$$

and the above. I chose these instead of others because it seemed to me that their number would be the smallest adequate number.

I assume further:

$$p . q . = :. (r) .\!\!\cdot\!\! . p . \supset . q \supset r : \supset . r \text{ Df}$$
$$p \lor q . = : p \supset q . \supset . q \text{ Df (disjunction)}$$
$$\sim p . = : p . \supset . (r) . r \quad \text{Df}$$

Concerning sense and meaning, I see nothing but difficulties which I cannot overcome. I explained the reasons why I cannot accept your view as a whole in the appendix to my book, and I still agree with what I there wrote. I believe that in spite of all its snowfields Mont Blanc itself is a component part of what is actually asserted in the proposition 'Mont Blanc is more than 4000 metres high'. We do not assert the thought, for this is a private psychological matter: we assert the object of the thought, and this is, to my mind, a certain complex (an objective proposition, one might say) in which Mont Blanc is itself a component part. If we do not admit this, then we get the conclusion that we know nothing at all about Mont Blanc. This is why for me the *meaning* of a proposition is not the true, but a certain complex which (in the given case) is true. In the case of a simple proper name like 'Socrates', I cannot distinguish between sense and meaning; I see only the idea, which is psychological, and the object. Or better: I do not admit the sense at all, but only the idea and the meaning. I see the difference between sense and meaning only in the case of complexes whose meaning is an object, e.g., the values of ordinary mathematical functions like $\xi + 1$, ξ^2, etc. But I admit that there are certain difficulties in this view. From what I have said about Mont Blanc you will see that I cannot accommodate the identity of all true propositions. For Mont Blanc is to my mind a component part of the proposition discussed above, but not of the proposition that all men are mortal. This alone proves that the two propositions are distinct from each other.

I have of course abandoned the definition of the equals sign which you criticize together with the whole view to which it belongs. I now assume

$$x = y . = . \varphi x \underset{\varphi}{\supset} \varphi y \text{ Df}$$

(To the objection that the equals sign to be defined occurs already as known I reply with Peano that '= . . . Df' counts for me as one symbol which does not express the same thing as '='. Definitions are not really part of the theory, but typographical stipulations. '= . . . Df' is not one of the primitive ideas of mathematics, but merely an expression of my will.)

I believe that $(4^2 - 3^2 = 7) \cdot = \cdot (7 = 7)$ is false. For to my mind, if we want to preserve identity, we must not replace one constituent part of a complex by another with the same meaning but a different sense. For I believe that the *sense* of '$4^2 - 3^2$' is essential to the proposition, i.e., the meaning of the component parts alone does not determine the proposition.

Yours sincerely,
Bertrand Russell

XV/20 [xxxvi/21]   FREGE to RUSSELL   9.6.1912

Jena
9 June 1912

Dear Colleague,

For a long time now it has been weighing on my conscience that I have not yet replied to your letter of 16 March.[1] I can well appreciate the great honour you did me by asking me to take part in the Mathematical Congress and to give a lecture there, and yet I cannot make up my mind to accept. I see that there are weighty reasons for my going to Cambridge, and yet I feel that there is something like an insuperable obstacle. And this is what makes it so difficult for me to answer your amiable letter. Please do not be angry at me for this.

And now I must further thank you cordially for the pleasure you and your co-author gave me by sending me the second volume of your *Principia Mathematica*. Because I was away on a trip, the parcel sat for a few weeks at the local customs office without my knowing its contents. Other occupations and lack of strength have kept me till now from taking more than a casual look at the book; but I hope to find time for a more thorough study. Once more my cordial thanks!

Yours sincerely,
G. Frege

XV/20. 1 The whereabouts of this letter is unknown.

XVI FREGE–STUMPF

Editor's Introduction

Carl Stumpf (1848–1936) obtained his doctorate at Göttingen in 1868 under H. Lotze, who was also one of Frege's teachers. Stumpf was a pupil of Brentano. In 1878 he became a professor of philosophy at Würzburg, and at the time of his correspondence with Frege he was a professor of philosophy at Prague. Stumpf's letter to Frege appears to be a reply to Frege's letter to Marty, Stumpf's colleague at Prague (cf. XII/1 above). Frege's letter may even have been addressed to Stumpf (cf. the editor's introduction to XII above).

XVI/1 [xl/1] STUMPF to FREGE 9.9.1882

Dippmannsdorf near Belzig
9 September 1882

Dear Colleague,

I should gladly comply with your wish to discuss the views you wrote about in your letter if I had your *Conceptual Notation* at hand, for unfortunately I do not quite remember it in detail, and the details are essential for a complete understanding of your views, and secondly, if I were not myself in the process of finally bringing to completion and into print an investigation I have been pursuing for seven years, which will claim all my time for several more months.[1] But I promise to answer you as soon as at all possible, and I also hope to be able to write something in a journal. (Could your new work perhaps be reviewed together with your *Conceptual Notation*?) In the meantime, let me say how pleased I am that you are working on logical problems, an area where there is such a great need for cooperation between mathematicians or scientists and philosophers. Incidentally, in my opinion, it is not just arithmetical and algebraic judgements which are analytic, but also geometrical ones – contrary to Riemann-Helmholtz and all English writers – and as soon as my work is finished, I shall move on to this topic.[2] Likewise, it seems to me (as it does to Brentano) that it is not just particular judgements which are existential, but also universal ones (the latter being negative existential judgements).[3]

XVI/1. 1 The investigation mentioned is Stumpf's *Tonpsychologie* I–II (Leipzig 1883 and 1890).

2 Starting from the investigations of B. Riemann and others on non-Euclidean geometry, H. von Helmholtz arrived at the view that the axioms of geometry are not only synthetic but also empirical judgements. J. S. Mill for example also held this view.

3 For example, 'All men are living things' can be reformulated (also in Frege's notation) as 'There is no man who is not a living thing'.

With regard to your work, to which I am looking forward with extra-ordinary interest, please do not take it amiss if I ask you whether it would not be appropriate to explain your line of thought first in ordinary language and then – perhaps separately on another occasion or in the very same book – in conceptual notation: I should think that this would make for a more favourable reception of *both* accounts. But I cannot, of course, judge this from a distance.[4]

Yours very sincerely,
C. Stumpf

---

4 Frege followed Stumpf's suggestion by publishing GLA before GGA, thus dispensing at first with his conceptual notation.

Editor's Introduction

Giovanni Vailati (1863–1909) studied under Peano and then collaborated with him from 1892 to 1899 as an assistant at the University of Turin. He later taught mathematics at various secondary and technical schools in Italy. Beyond his more narrowly scientific and teaching activities, Vailati engaged in wide-ranging activities as an author and educational reformer, which brought him in contact, through correspondence, travels, and conferences, with many well-known contemporaries, among them Franz Brentano, Ernst Mach, Benedetto Croce and Georges Sorel. In 1905 he became a member of a commission for the reform of secondary education in Italy, and in the following year he visited Germany as a member of the commission and met Frege in Jena. The *Scritti di G. Vailati* (Leipzig and Florence 1911) as well the letters edited by G. Lanaro in G. Vailati, *Epistolario 1891–1909* (Turin 1971), document his wide-ranging activities. A short biography and an assessment of Vailati's contributions to science are contained in Luigi Einaudi, 'Ricordo di Giovanni Vailati', in the volume of correspondence cited above, pp. 19–26.

Vailati wrote to Frege in French.

XVII/1 [xliii/1]   Vailati to Frege   17.3.1904

Como
17 March 1904

My dear Professor,

I have just read with the greatest interest the short works you were kind enough to send me, and I venture to set down at once some of the reflections they have suggested to me, especially with reference to your critical remarks on Mr Hilbert's work, *On the Foundations of Geometry*.[1]

I admit that you are quite right in regarding Hilbert's exposition, as it relates to the axioms, as absolutely *incoherent*. His claim to characterize them as expressing 'fundamental facts of our intuition' plainly contradicts the use he makes of them and the sense (or better, *lack of sense*) which he in effect attributes to them in his treatise.[2] But I believe that if only Mr Hilbert

---

XVII/1. 1 In a letter to Giovanni Vacca dated 22.3.1904, Vailati mentions that Frege sent him his paper 'What is a Function?' ⟨20⟩ as well as 'two further short publications' containing a critique, 'not a very felicitous one', of Hilbert's *Grundlagen der Geometrie*. The publications in question are evidently Frege's series of essays GLG III 1–3 or the first two parts of it.

2 Vailati is referring to sect. 1 of *Grundlagen der Geometrie* (1899), where Hilbert writes: 'The axioms of geometry are divided into five groups; every one of these groups expresses certain fundamental facts of our intuition which belong together.'

could make up his mind to renounce his opinion that the axioms represent the 'fundamental facts of intuition' (an opinion whose source is perhaps to be sought in his irrational devotion to Kantian philosophical jargon, which is still deplorably popular among writers on the philosophy of science), all the rest of his exposition could be given an irreproachable form:

When he says that a given group of axioms defines a 'relation' (for example 'between') or a class of objects (for example 'points'), he does not sufficiently distinguish the different nature of the definition in the two cases:[3]

(1) To define a *'relation'* by means of axioms is to enunciate a *condition* (or functional equation) or several conditions which are supposedly satisfied by the relation being defined.

(2) To define a *class of objects* by means of axioms is to characterize it as constituted by objects *between which* one can establish the 'relations' that satisfy the conditions enunciated in the axioms.

In the first case we have, so to speak, a *direct* definition, and in the second an *indirect* definition (that is, a definition of a *class of objects* by means of other definitions, that is, by means of the definitions of the relations that can supposedly be established between them: as when 'quantity' is defined as any object belonging to a class *between whose members* [or constituents] there exist the relations $<, >, =$, which are defined by means of their well-known formal properties).[4] The example you cite (p. 370): 'Any God is omnipotent' etc. is not an example of a definition of a *relation*, and this is why it is objectionable in my opinion.[5]

When Hilbert enunciates the axiom: 'For any three points [on a line] *there is* always one [and only one] point which lies *between* the two others', he must be understood to be enunciating, not a characteristic mark of 'points', but a characteristic mark of the relation 'between';[6] for otherwise he would be committed to the absurd belief (denounced already by Aristotle, as I showed in the pamphlet *Aristotle on Definition*, p. 14, which I had the honour of sending you some time ago)[7] that the existence of some 'thing' could form part of its 'definition' – (τὸ δ'εἶναι οὐκ (ἐστίν) οὐσία οὐδενί· οὐ γὰρ γένος τὸ ὄν).[8]

---

3 *Grundlagen der Geometrie*, sect. 3; of also IV/3 above, n. 3.

4 Underneath the words *entre les membres de laquelle* ('between whose members') Vailati wrote (in English) 'between whose constituents'.

5 GLG II, p. 370 [E.T. p. 32].

6 The quotation has been completed according to Hilbert's text: Vailati omits the words in brackets.

7 *Sulla definizione in Aristotele.* There seems to be no publication with this title. Vailati may possibly be thinking of his essay, 'La teoria aristotelica della definizione', *Rivista di Filosofia e scienze affini* (Padua), vol. II, yr. V. No. 5 (Nov.–Dec. 1911), pp. 485–96.

8 *Posterior Analytics* II 7, 92b, 13ff. The insertion in brackets is Vailati's.

Please excuse my haste, which obliges me to neglect the formalities. I should like to prolong my conversation with you by discussing other important points in your writings, which I find among the most stimulating writings I have ever read on these topics.

Can a non-German belong to the German Mathematicians' Association (or is it a German-mathematicians' association)? If possible, I should like to apply for membership.

Yours sincerely,
G. Vailati

XVIII FREGE–ZSIGMONDY

Editor's Introduction

Karl Zsigmondy (1867–1925) was professor of mathematics at the Vienna Polytechnic at the time Frege composed the following letter (after 1918). As a student in Vienna (1886–90) Zsigmondy had already been especially interested in algebra and number theory. In the summer semester 1891 and in the winter semester 1891–2, he studied in Berlin under L. Kronecker, who seems to have had a lasting influence on him. Subsequently, Zsigmondy devoted himself exclusively to number theory. In 1918, after being elected rector, he delivered an inaugural address 'Zum Wesen des Zahlbegriffs und der Mathematik'.[1] He probably sent an offprint of the address to Frege, to which Frege replied in the following letter: Frege's exposition makes special reference to the definition of a cardinal number contained in the address.

XVIII/1[xlvi/1] FREGE to ZSIGMONDY   undated

As you probably know, I have made many efforts to get clear about what we mean by the word 'number'.[1] Perhaps you also know that these efforts seem to have been a complete failure. This has acted as a constant stimulus which would not let the question rest inside me. It continued to operate in me even though I had officially given up my efforts in the matter. And to my own surprise, this work, which went on in me independently of my will, suddenly cast a full light over the question.

Where does the difficulty really lie? I believe it goes back to the time when we first became acquainted with the word 'number'. Perhaps there surges up

---

1 The address appeared in *Bericht über die feierliche Inauguration des für das Studienjahr 1918/19 gewählten Rector Magnificus Dr. Karl Zsigmondy o.ö. Professors der Mathematik am 26. Oktober 1918* (Vienna 1918), pp. 41–78.

XVIII/1. 1 In his inaugural address, Zsigmondy describes the formation of the concept of 'cardinal number' as follows:

> The concept of equivalence is of fundamental significance for set theory, for the reason that sets can be divided into classes by means of it; in particular, since two sets equivalent to a third are also equivalent to each other, all sets of equal potency or valency can be regarded as belonging to one and the same class. What is common to all equivalent sets united in a class, i.e., to all sets related to one another by a one-one relation, the invariant feature, as it were, that characterizes their class, is conceived as a newly formed concept, as an idea subsisting by itself, and is called the cardinal number of every one of the sets belonging to the class in question.
>
> We thus arrive at the idea of the cardinal number of a given set by abstracting from the characteristics and, as the case may be, also from the context, or from the arrangement of its elements.

Cf. op. cit., pp. 47ff.

in our memory a heap of peas which a child is trying to count. 'A number is a heap' – this is perhaps the first proposition to occur to us here. If we say instead, 'A number is a heap of things', this already sounds more scholarly; but does not every heap consist of things? The addendum 'of things' is therefore unnecessary. It sounds even more scholarly when a professor of mathematics announces 'A number is a series of objects of the same kind'. But whether the peas form a heap or are arranged in a series makes no difference to their number. The word 'of the same kind' attracts our attention, and yet it does not really say anything at all. Different things always differ in something and are therefore of different kinds, and on the other hand they almost always agree in something and are therefore of the same kind. If we take things as our basis, we must not intermix what belongs to the way someone conceives things but is not proper to the things themselves. After stripping off all unnecessary accretions, we accordingly get back to the proposition 'A number is a heap', where we also count a series of things as a heap.

Number words and number signs are called by these names only because they are supposed to serve to designate numbers. If numbers are heaps or series, it is to be expected that a number word serves to designate a heap. Now what heap does the word 'five' designate? No one can say. Nor can anyone say where the heap named 'five' is located. And where is the heap named 'one'? Or where is the series of objects of the same kind named 'one'? None of the number words 'one', 'two', 'three' and none of the number signs '1', '2', '3' designates a heap or a series. There are therefore numbers which are not heaps or series. Perhaps no number at all is a heap. We thus give up the propositions 'A number is a heap' and 'A number is a series of objects of the same kind' as untenable.

Yet it is at first difficult to think that a number has nothing to do with a heap. It almost seems that, while a number is not a heap, it is something in a heap. This would explain that we think we find the same number in different heaps, as we can recognize the same colour in different things. When, then, do we find the same number in different heaps? We may think of the following situation. Say we have a heap of peas and a heap of beans. Now if we succeed in pairing the peas of the first heap and the beans of the second in such a way that each pair consists of a pea and a bean and none of the peas and none of the beans is left over without a partner, then the first heap has the same number as the second. The number thus appears as a property of heaps.

But what is at issue is the use of number words and numbers signs in the science of mathematics. And there we find propositions like

'six is an even number',
'five is a prime number',

propositions that express the subsumption of an object (six, five) under a concept (even number, prime number), just as the proposition

'Sirius is a fixed star'

expresses the subsumption of an object (Sirius) under a concept (fixed star). In their scientific use in mathematics, number words are therefore proper names like 'Sirius' or 'Africa'. What a proper name designates is an object. Mathematics therefore regards numbers as objects, not as properties. It uses number words substantivally, not predicatively.* This gives rise to the notion that to every heap there belongs a number, not as a property but as an object.

Any number belongs in general to more than one heap. This yields at once a division of heaps into classes, where all heaps that have been assigned the same number are assigned to the same class. Thus to any number there corresponds a class of heaps and to any class of heaps a number. To different classes of heaps correspond different numbers and to different numbers different classes of heaps. Now what more do we really know about numbers than that we can recognize the same number again and that we can distinguish different numbers. The same holds for classes of heaps. Now this strongly suggests that we say: Classes of heaps are numbers, and numbers are classes of heaps. We thus drop the distinction between numbers and classes of heaps entirely. Does this not give us everything we need?

---

* This distinguishes the scientific use of number words in mathematics from their unscientific use in daily life, where the use of 'four horses' is similar to that of 'big horses', as if 'four' gave us a property of an individual horse like 'big'; and yet, while in the case of big horses every individual is big, in the case of four horses not every individual is four.

APPENDIX

Gottlob Frege[1]
by Philip E. B. Jourdain

[A chapter from Philip E. B. Jourdain, 'The development of the theories of mathematical logic and the principles of mathematics' in *The Quarterly Journal of Pure and Applied Mathematics* 43 (1912), pp. 237–69. All notes are Jourdain's own or his report of Frege. Small corrections of text and symbolic matter have been made without special indication. Some assistance in following the references is given in square brackets. Jourdain's promise, at the end, of a further analysis of GGA was never fulfilled.]

Abbreviations used by Jourdain

Bs. – *Begriffsschrift.*	*B.P.* – *Über d. Bs. des Herrn Peano.*
Gl. – *Grundlagen der Arithmetik.*	*R.d.M.* – *Revue de Mathématiques.*
S.u.B. – *Ueber Sinn u. Bedeutung.*	*R.Pr* – Russell's *Principles.*
F.u.B. – *Function u. Begriff.*	*A.J.M.* – *Amer. J. of Math.*
Gg. – *Grundgesetze der Arithmetik.*	*R.M.M.* – *Rev. de Mét. et de Morale.*

— 1 —

The increasing tendency towards more rigorous proofs in mathematics, the more accurate determination of limits of validity, and, for these purposes, a sharp definition of concepts, showed itself during the nineteenth century, and, after the concepts of function, of the limits and continuity of a function, of the infinite, and of negative and irrational numbers had shown themselves to need more accurate definitions, this path led to exacter investigations in the foundations of arithmetic.[2]

Frege proposed to himself the question as to whether arithmetical judgements can be proved in a purely logical manner or must rest ultimately on facts of experience; and, consequently, began by investigating how far one could go in arithmetic by conclusions which rest merely on the laws of general logic. He tried first of all to reduce the concept of arrangement in a series to that of *logical* sequence, in order to proceed from this point to the number-concept. In order that nothing intuitional should intrude unremarked, all ultimately depended upon the unbrokenness of the chain of inferences; and ordinary language was found to be unequal to the accuracy required for this purpose. Hence arose the 'Begriffsschrift', which was described and shown in use in a small book published in 1879.[3] The fundamental thought of the

---

1 Friedrich Ludwig Gottlob Frege was born on 8 November, 1848, at Wismar, and from 1874 on has taught at the University of Jena.

Professor Frege has most kindly read this paper in manuscript, and added the notes marked '(F. 1910)'.

2 Frege, *Gl.*, pp. 1–2. Cf. article 13 below.

3 *Begriffsschrift, eine der arithmetischen nachgebildete Formelsprache des reinen Denkens* (Halle a. S. 1879), x, 88 pages (preface dated 18 December, 1878). The statement of the origin of this ideography, which is given above, was on p. iv. Cf. also *Gl.*, referred to below, pp. 1–5, 102–4; and *B.P.*, pp. 362–3 (see article 13 below).

Begriffsschrift[4] was the transference[5] of the distinction of variables and constants, which was well known, but not thoroughly carried out,[6] in algebra and analysis, to the wider domain of pure thought in general. Thus, Frege divided all the signs which he used into:(1) letters, 'by which one can represent to oneself different things', like those in the generally valid theorem

$$(a + b)c = ac + bc,$$

and which serve principally to express *generality*; (2) signs which have quite a definite sense, like $+, -, \sqrt{}, 0, 1, 2.$[7]

---

4 Venn (*Symbolic Logic*, 1st ed. (1881), p. 415: cf. 2nd ed. (1894), pp. 493–4) has remarked that Frege worked out his scheme 'in entire ignorance that anything of the kind had been achieved before'. Though, indeed, Leibniz, alone of Frege's predecessors, was mentioned by name in the *Bs.* (p. v.), that Frege was not unacquainted with the work of others on the calculus of logic may, I think, be inferred from a passage on p. iv.: 'The modelling upon the arithmetical formula language to which I alluded in the title refers more to the fundamental ideas than to the detailed configuration. Nothing was further from my mind than all those efforts to create an artificial similarity by conceiving a concept as the sum of its characteristic marks.'

Cf. the references to Boole's work towards the end of article 12 below.

The difference between the aims of Frege, on the one hand, and all the other previous cultivators of the calculus of logic, on the other, is dealt with further in the text below.

5 *Bs.*, p. 1. Cf. *F.u.B.*, pp. 12, 30–31.

6 Think, for example, of *l*, log, sin, Lim (Frege's examples, *Bs.*, p. 1n).

7 It must be remembered that (cf. Frege, *Was ist eine Funktion? Festschrift, Ludwig Boltzmann gewidmet zum sechzigsten Geburtstage* (Leipzig 1904), pp. 656–6) what are called 'variables' in mathematics have nothing to do in themselves with time. When people speak of a 'variable magnitude' – for instance, the length of a rod under various conditions – what they mean is that, though the length is a definite number at each instant, at different instants the length may differ. The number of millimetres in the length of this rod denotes no number unless a particular instant is given; the phrase is a *functional* phrase, and the function is what Russell called a *denoting* function.

Also B. Russell (*Mind*, N.S., xvii (1908), p. 240 [*B. Russell, Mr Haldane on Infinity*, l.c., pp. 238–42]) has expressed it: 'A variable is (represented by) a symbol which is to have (or rather represent) one of a certain set of values, without its being decided which one. It does not have first one value of a set and then another; it has at all times *some* value of the set, where, so long as we do not replace the variable by a constant, the "some" remains unspecified.' We have added alterations which appear to us to make the description more exact.

On the word 'variable', Frege has supplied the note:

Would it not be well to omit this word, since it is hardly possible to explain it properly. On Russell's explanation, the first question that arises is as to the meaning of the phrase: 'a symbol has a value'. Is the relation of a sign to its signification meant by this? In that case, the symbol can only have one value, since the sign must indicate uniquely, and it must be determined which value the sign is to have. Then the variable would be a sign. If, now, you write: 'A variable is represented by a symbol which is to have one of a certain set of values', the last fault is removed thereby; but what is the case then? The symbol represents, in the first place, the variable, and, in the second, one of a certain set of values without it being determined which. Accordingly, it seems best to leave the word 'symbol' out of the explanation. The question as to what a variable is, besides, to be answered independently of the question as to by which symbol a variable is to be represented. So we come to the explanation: 'A variable is one of a certain set of values without it being decided which ones'. But this last addition does not give any closer determination, and to belong to a certain set of values properly means to fall under a certain concept; for this set can only be determined by the giving of the properties that an object must have in order to belong to this set; that is to say, the set of values becomes the extension of a concept. But we can, now, for every object, give a set to which it belongs, so that also the requirement that something is to be one of a certain set of values determines nothing. The best thing to do is to remain by the statement that the Latin letters serve to confer on a theorem generality of the content. And it is better not to use the expression 'variable' at all, since at bottom we can neither say of a sign nor of what it expresses or denotes, that it is variable or a variable – at least not in a sense which can be used in mathematics or logic. On the other hand, perhaps someone may emphasize that in '$(2 + x)(3 + x)$' the letter 'x' does not serve to confer on a theorem generality of the content. But in connexion with a proof such a formula will always appear as a part of a theorem, which may also partly consist of words, and 'x' will, in such a connexion, always serve to confer on a theorem generality of the content. Now, it seems to me to be unfortunate to limit the values permissible to this letter to a set. We can, indeed, always add the condition that x belongs to this set, and then allow that limitation to drop. If an object Δ does not belong to that set, that condition is not fulfilled for it, and if we substitute 'Δ' for 'x' in the whole theorem, we get a true theorem. I would not say of a letter that it has a *meaning* if it serves to confer on a theorem generality of content. We can replace the letter by the proper name 'Δ' of an object Δ; but this Δ cannot anyhow be regarded as *meaning* of the letter; for it is not more closely allied with the letter than with any other object. Also, generality cannot be regarded as meaning of the Latin letter; for it is not something independent which could be added to the theorem which is already complete without it. I would not, then, say 'terms whose meaning is indeterminate' or 'signs have variable meanings'. In this case the signs have no denotations at all (F., 1910).

In universal algebra, as Russell has pointed out (*Mind*, N.S., xii (1903) [*B. Russell, Recent Work on the Philosophy of Leibniz*, l.c., pp. 177–201], p. 187), our signs of operation have variable meanings. However, he went on, it is not the logically first of all mathematical subjects, for it employs deduction and the logical kinds of synthesis, which are explicitly dealt with in the logical calculus.

The matter may be stated thus: In every proposition, when fully stated, there must be cônstants, i.e., terms whose meaning is not in any degree indeterminate. When we turn our symbols of operation into variables, we do not thereby remove all constants from our propositions, for the formal laws to which our operations are to be subjected will require constants for their statement. I have succeeded in reducing the number of indefinable terms employed in pure mathematics (including geometry) to eight (a number which may be capable of further diminution), by means of which every notion occurring throughout the whole science can be defined. Thus all mathematics is merely the study of these eight notions; and the Logical Calculus is a name for the more elementary parts of this study. We have here precisely such a development as Leibniz desired to give to all subjects, with the difference, due to the fact that all propositions are synthetic, that the indemonstrable axioms of mathematics, instead of being one, appear to number about twenty.

And to this Russell added the note:

The only ground, in Symbolic Logic, for regarding an axiom as indemonstrable is, in general, that it is undemonstrated; hence there is always hope of reducing the number. We cannot apply the method by which, for example, the axiom of parallels has been shown to be indemonstrable, of supposing our axiom false; for all our axioms are concerned with the principles of deduction, so that, if any one of them be true, the consequences which might seem to follow from denying it do not follow as a matter of fact. Thus from the hypothesis that a true principle of deduction is false, valid inference is impossible.

On the sentence: 'In universal algebra ... our signs of operation have variable meanings', Frege remarked in 1910:

If, for example, we wish to investigate what follows from the laws $a + (b + c) = (a + b) + c$ and $(a + b) + c = (a + c) + b$, quite independently of the usual denotation of the sign of addition, one ought wholly to avoid the word 'addition' and the sign '+' and express the laws thus: $f\{a, f(b, c)\} = f\{f(a, b), c\}$ and $f\{f(a, b), c\} = f\{f(a, c), b\}$. Now the letter '$f$' serves to make the consideration general. If we wish to understand signs by 'variables' and 'constants', variables will be signs which indicate indefinitely, and constants will be denoting signs. It is to be wished that letters alone were used for the former and never letters for the latter. Just as little as we use the number-signs '2' and '3' like letters, in order to confer on a theorem generality of the content, ought one to use the signs '+', '−', and '$\sqrt{\ }$' like letters. It is customary in arithmetic to use them as denoting signs, and one ought to remain by that. On the other hand, it is to be wished that a letter were never used as a proper name; for example, the basis of the system of natural logarithms.

On the sentence: 'In every proposition ... indeterminate', Frege (1910) remarked:

Often I am in doubt as to whether I ought to understand by the word 'proposition' an expression of a thought formed out of audible or visible signs, or a thought. Here is said that, in a *proposition*, *terms* must be contained, and that there *terms* have *meanings*. Hence we may conclude that the *terms* are signs, and hence, too, the *proposition* is to be an expression of a thought and not a thought itself. Accordingly, we must here understand by 'constants' denoting signs. But the assumption that *terms* are to be signs does not agree with the expressions that follow it: 'indefinable terms, by means of which every notion ... can be defined'. Accordingly, definability seems to come into question both with *notion* and *term*, and this is hardly to be reconciled with the assumption that a *term* is to be a sign, for *notion* is obviously not a sign.

On 'at all times *some* value', Frege (1910) remarked:

That Russell here uses the word 'times' appears to me unfortunate, since the time has nothing to do with it. If we translate here 'time' by the German 'Mal', it is questionable how and whereby the various times (*Male*) differ from one another. The word 'times' is not quite clear to me.

On 'Thus from the hypothesis ... impossible', Frege remarked in 1910:

From false premises nothing at all can be concluded. A mere thought, which is not recognized as true, cannot be a premise. Only after a thought has been recognized by me as true, can it be a premise for me. Mere hypotheses cannot be used as premises. I can, indeed, investigate what consequences result from the supposition that A is true without having recognized the truth of A; but the result will then contain the condition *if A is true*. But we say thereby that A is not a premise, for true premises do not occur in the concluding judgement. Under circumstances we can, by means of a chain of conclusions, obtain a concluding judgement of the form

— 2 —

We have seen that arithmetic was the starting-point of the road which led Frege to his ideography;[8] and the aim of his ideography was not to provide a means of dealing systematically and rapidly with complicated logical questions, but to enable one to settle the question as to the empirical or merely logical basis of a branch of knowledge – in this case, arithmetic

When we raise the question as to the foundation of a truth, the answer which – unlike that given by the recounting of the historical genesis and development of our knowledge of the truth in question – is connected with its inner being,[9] consists in carrying out its proof purely logically, if that is possible, or, if it is not, in reducing it to the facts of experience on which the proof rests.[10] The firmest proof is obviously a purely logical one, which, abstracting as it does from the special nature (*Beschaffenheit*) of the things, is founded alone on the laws on which all knowledge rests. Certainly a proposition may be capable of logical proof and yet could never, without sense-perception, enter into our consciousness. Indeed, this seems to be the case with every judgement, since no mental development without sense-perception appears possible. Thus, it is not the psychological origin, but the completest manner of proof, that brings about the division of the class of all truths which need founding into (*a*) those which can be proved purely logically, and (*b*) those whose proof rests on facts of experience.[11] Frege compared [12] the relation of his ideography to ordinary language to that of the microscope to the eye: his ideography was a means of investigation devised for definite scientific ends, and it is no argument against it to point out that for certain other ends it is incomplete or inconvenient.

And the very important ends for which Frege's ideography was designed were more or less overlooked by Venn,[13] Schröder, and Peano, who criticized principally the cumbrousness of Frege's notation. This cumbrousness is a fact, and may, as Russell[14] has shown, be avoided to a great extent; but far more important than the awkwardness of the form of many of the symbols is the subtler and more profound analysis of the ideas of logic, the fact that the residuum of this analysis is denoted

---

Here A, B, and Γ do not appear as premises of the method of conclusion, but as conditions in the concluding judgement. We can free this judgement from the conditions only by means of the premises $\vdash A$, $\vdash B$, $\vdash \Gamma$, and these are not hypothesis, since their signs contain the sign of assertion.

The indemonstrability of the axiom of parallels cannot be proved. If we do this apparently, we use the word 'axiom' in a sense quite different from that which is handed down to us. Cf. my essays 'Über die Grundlagen der Geometrie,' in Bd. xv of the *Jahresber. der D.M.-V.*

8 Cf. *Bs.*, pp. iv, viii. Cf. *Gl.*, pp. 3–5; *Gg.*, i, pp. viii–ix, l.

9 Cf. the reference to Leibniz in *Gl.*, p. 23. See *Gg.*, i, pp. xiv–xxvi on this and on psychology in logic.

10 Frege's view (F., 1910) was that 'the truths of geometry, in particular the axioms, are not facts of experience, at least if by that is meant that they are founded on sense-perceptions'.

11 *Bs.*, pp. iii–iv.

12 ibid., p. v. Cf. pp. vi–viii.

13 op. cit., 1st ed., p. 415; 2nd ed., pp. 493–4.

14 *A.J.M.*, xxviii and xxx (1906 and 1908) [*The theory of implication*, l.c., pp. 159–202, *Mathematical logic as based on the theory of types*, l.c., pp. 222–62].

by a class of trùly constant symbols, and the more perfect avoidance of ambiguity and implicit assumptions, which form the prominent characteristics of Frege's work.

— 3—

In the *Begriffsschrift* a judgement was expressed by the help of the sign

$$\vdash,$$

which stood on the left of the sign or combination of signs which expressed the content (*Inhalt*) of the judgement. The horizontal dash (the *Inhaltsstrich*, as opposed to the vertical *Urtheilsstrich*) connected the signs following it to a whole, and the assertion expressed by the judgement-stroke referred to this whole. If the judgement-stroke is omitted the judgement is to change into a mere combination of presentations (*Vorstellungsverbindung*),[15] of which the writer is supposed to express no decision as to the truth or otherwise of the judgement. If, for example, the above sign followed by the letter '*A*' says (*besagt*[16]) 'opposite magnetic poles attract one another', then the combination of signs without the vertical stroke at the left end does not express this judgement, but merely serves to call up in the mind of the reader the presentation of the mutual atrraction of opposite poles.[17]

We use in this case the circumlocution 'the circumstance that ...' or 'the proposition that ...'.[18] Not every content can become a judgement by prefixing to its sign a sign of assertion; thus, the presentation 'house' cannot. Accordingly, Frege distinguished *judicable* (*beurtheilbare*) and *non-judicable* contents.[19]

---

15 Instead of this word I now say more simply 'Gedanke'. The word 'Vorstellung' is used, now in a psychological, now in a logical sense. Since obscurities arise from this, I have decided not to use it at all in Logic.

    One must be able to express a thought without asserting its truth. If we wish to represent (*hinstellen*) a thought as false, we must first express it without assertion, then add the denial, and then represent the so expressed new thought as true. We cannot express correctly a hypothetical thought-combination if we cannot express thoughts without representing them as true; for, in the hypothetical combination, neither the thought appearing as condition nor that appearing as consequent is asserted (F., 1910).

16 This word was suggested to me by Frege in October 1910. Frege said: 'A proposition (*Satz*) expresses (*ausdrückt*) a thought and signifies (*bezeichnet*) its truth-value. Of a judgement one can neither properly say that it is signified, nor that it is expressed. In it, however, we have a thought, and this can be expressed; but we have more, namely, the recognition of the truth of this thought.'

    Frege also substituted for that part of the above sentence beginning: 'but merely serves ...' the words: 'but only expresses the thought that opposite poles attract one another'.

17 Cf. Russell, *Principles*, p. 35, and *A.J.M.*, xxviii, pp. 161, 176; Frege, *F.u.B.*, pp. 21–2, and *Gg.*, i, p. 9.

18 Instead of 'circumstance' (*Umstand*) and 'proposition' (*Satz*), Frege suggested 'thought' (F., 1910).

19 *Bs.*, p. 2. An example of another kind of a non-judicable content was given on p. 64 (ibid.), formula 81. Russell (*Principles*, p. 519) apparently did not observe that Frege, by demanding that all his contents should be judicable (should be unasserted propositions), introduced an hypothesis which applied to all the definitions and theorems of the *Begriffsschrift*, and which he did not avoid until 1891 (see 4th note to article 4 below, and article 23 of the section on Peano below).

In the representation of a judgement in the *Begriffsschrift*, there was no distinction of subject and predicate;[20] for two judgements which only differ in the place of certain words in the series of words denoting them only differ in psychological intention, but the same conclusions follow from both the judgements in connexion with certain definite other ones. Frege said that the two judgements have, in this case, the same 'begriffliche Inhalt', and remarked that, in not distinguishing between subject and predicate, the *Begriffsschrift* follows the example of mathematical symbolism. 'A language can be imagined in which the sentence: "Archimedes was slain at the conquest of Syracuse" is expressed: "the violent death of Archimedes at the conquest of Syracuse is a fact". Here we can indeed, if we wish, distinguish subject and predicate, but the subject contains the whole content, and the sole aim of the predicate is to make it a judgement. Such a language, which has one and the same predicate – namely, " is a fact" – for all judgements, is the *Begriffsschrift*, and the sign |— is its common predicate for all judgements.'[21] Frege remarked[22] that, in his first sketch of a symbolic language, he allowed himself to be guided by the example of ordinary language, and made judgements to be composed of subject and predicate; but he soon found that this was only a hindrance for his special ends, and only led to useless prolixity. Also, the distinction of judgements into categorical, hypothetical and disjunctive judgements appeared to Frege[23] to be merely of grammatical significance.

— 4 —

If '*A*' and '*B*' denote judicable contents, there are four possibilities obtained by asserting and denying each in turn. Frege[24] denoted by

the judgement that that one of these possibilities in which *A* is denied and *B* asserted does not present itself. Thus, if we call the vertical line which connects the two horizontal lines the 'line of condition' (*Bedingungsstrich*) and the relation between *B* and *A* that we have just defined 'conditionality' (*Bedingtheit*) or generalized implication of *A* by *B*,[25] we can say that a true content is implied by a true or a false content, and that a false content implies either a true or a false content.[26] The essential point is that a true content can only imply a true content; and, if we do not know whether *A* and *B* are to be asserted or denied, the implication may be translated: 'if *B* then *A*'. The causal connexion, which lies in the word 'if' is, however, not expressed by the sign of implication, and accordingly we can speak of the relation of implication just defined as a generalization of the 'implication' of common language.[27]

---

21 ibid., pp. 3–4.   22 ibid., p. 4.   23 ibid.   24 ibid., p. 5.
25 The essential property that we require of implication is that what is implied by a true proposition is true, so that it cannot be the case that *B* is true and *A* not true.
26 ibid., pp. 5–6.
27 Cf. Venn, op. cit., 1st ed., p. 341n; 2nd ed., p. 242; the above section on MacColl; and Russell, op. cit., pp. 14–15, 518–19. In the last passage mentioned (on implication with Frege), Russell remarked that Frege's relation holds when *B* is not a proposition at all, whatever *A* may be; but this, as we have already pointed out, seems to be a later introduction of Frege's, with the purpose of avoiding the necessity for the hypothesis '*A* and *B* are propositions.' Cf. also Russell, *A.J.M.*, xxviii (1906), pp. 161–2; xxx (1908), pp. 244–5.

The extension of Frege's symbolism[28] to the representation of 'Γ implies that "B implies A"' (that it cannot be that A is denied, B and Γ asserted), and '"B implies A" implies Γ' (either Γ is true or B is true and A is false[29]) is obvious. We have only to remember that, in the symbol for 'B implies A', the part of the upper horizontal line on the left of the line of condition is the content-line of the implication, and to this every sign which refers to the total content of the expression is applied.

From the above explanation of implication results, according to Frege,[30] that, from the judgements represented by (1) and $\vdash B$, the new judgement $\vdash A$ follows. And, from the judgement 'Γ implies that "B implies A"', and the assertions of Γ and of B, the assertion of A follows.[31]

Of all the Aristotelian manners of conclusion, Frege[32] only used this one, at least in all cases in which from more than a single judgement a new one is derived. For we can state the truth in another manner of conclusion in a judgement of the form: If M holds, and if N holds, then A also holds; in signs:

$$\begin{array}{l} \rule{1cm}{0.4pt} \\[-6pt] \quad\rule{0.4pt}{1.2cm}\!\!\!\rule{0.8cm}{0.4pt}\,A \\[-4pt] \quad\qquad\rule{0.4pt}{0.6cm}\,M \\[-2pt] \qquad\qquad\rule{0.4pt}{0.4pt} N. \end{array}$$

From this judgement and $\vdash N$ and $\vdash M$ there follows $\vdash A$, as above.

— 5 —

The denial of A, Frege expressed by writing a short vertical line below the content-line, half-way between the 'A' and the sign of assertion;[33] and of the line of negation was then placed at different points of the schemes already treated, and the results interpreted; and also 'or' and 'and' were represented by implications and negations.[34] Frege remarked explicitly[35] that 'the words "or", "and", "neither-nor" evidently only here come into consideration in so far as they connect *judicable* contents'.

---

28 *Bs.*, pp. 6–7.
29 On p. 7 (ibid.), Frege had, by mistake, the interpretation: the case where B is true and A and Γ false cannot be. Schröder (see article 11 below) pointed out the mistake; the correct interpretation may be written (in Schröder's symbols)

$$\overline{\overline{B\overline{A}\Gamma}} = B\overline{A} + \Gamma$$

while Frege's interpretation is

$$\overline{\overline{BA\Gamma}} = \overline{B} + A + \Gamma.$$

30 *Bs.*, p. 7. In fact, the assertion of 'B implies A', and of B, leaves only the possibility 'A is true' open.
31 ibid., pp. 8–9. We distinguish between 'implication', which is unasserted, and an *asserted* implication or *judgement* of the implication. Cf. Russell, *Principles*, p. 35.
32 *Bs.*, pp. 9–10.
33 This line of negation was an inverted 1 in the *Begriffsschrift*, but was joined on to the content-line in the *Grundgesetze* of 1893.
34 *Bs.*, pp. 10–13; *F.u.B.*, p. 23.
35 *Bs.*, p. 13; Russell, *A.J.M.*, xxviii, p. 162.

— 6 —

Whereas implication and denial referred to contents, and signs were hitherto merely representatives of their contents, when two signs are connected, by the sign of equality (*Inhaltsgleichheit*)³⁶, ≡, the circumstance that two names have the same content is denoted. In geometry, for example, we often arrive by means of different manners of determination (constructions) at the same content (point); and hence the different names for the same content are not always merely an unimportant matter of form, but they concern the essence of the matter itself if they are connected with different manners of determination. Further, it is sometimes convenient to introduce a sign to replace a long expression, and this also needs the sign of equality. The assertion of $A \equiv B$ means, then, that the sign 'A' and the sign 'B' have the same content, so that we can put everywhere 'B' in the place of 'A' and inversely.³⁷

— 7³⁸ —

Suppose that in an expression about an entity represented by 'A', we have replaced the sign 'A' by signs (one after another) denoting other entities, the expression becomes what mathematicians would call 'variable owing to the variability of A'; and there appears a 'constant part which represents the totality of the relations' and 'the sign which is thought of as replaceable by others, and which denotes the object which finds itself in these relations'; the first part Frege³⁹ called 'function', the second 'its argument'.⁴⁰ Generally,⁴¹ 'if in an expression, whose content need not be judicable, a simple or compound sign occurs at one or many places, and we think it as replaced at some or all of these places by something else, but everywhere by the same thing, we call that part of the expression which appears in it as invariable, *function*, and the replaceable part *its argument*'.⁴² We get functions of two and more

---

36 *Bs.*, pp. 13–15.
37 ibid., p. 15. That the sign of equality indicates a relation between *names*, Frege here held in common with Venn (op. cit., 1st ed., p. 64; 2nd ed., p. 70), Peano (see the section on Peano, below), and many others; but later, by distinguishing between the *denotation* and *meaning* of a sign (see below), he held that an equation expresses identity of denotation with diversity of meaning.
  On the ideas associated with '≡' and '=' see below, and *Gg.*, i, p. ix.
38 This article ('Die Function') is from *Bs.*, pp. 15–19; cf. *F.u.B.* and *Gg.*, i, pp. 5–25.
39 *Bs.*, p. 15.
40 Notice that here Frege, although he was careful to distinguish the sign 'A' from what it denoted, spoke of the *expression* as variable, and defined the function as a certain part of the *expression*. The correction of this error, upon an analogous one to which Frege insisted so strongly (from 1884 on; see below) in his criticisms of formalist theories of arithmetic and the usual mathematicians' definition of function and variable, is one of the advances made in the *Function und Begriff* of 1891.
  The distinction of function and argument referred to has nothing to do with the conceptual content, but is merely a matter of the mental grasping (*Auffassung*); *Bs.*, pp. 15–16.
41 *Bs.*, p. 16.
42 On p. 17 (ibid.), Frege cautioned against the mistake, to which verbal usage leads, of thinking that in the propositions 'the number 20 is representable as the sum of four squares', it is possible to replace the argument 'the number 20' by the concept of another rank (*Rang*) 'every positive integer'.

arguments by thinking a sign (in a function), hitherto regarded as not replaceable, replaceable at all or some of the places where it occurs;[43] and indefinite functions of the argument were denoted as in ordinary analysis, but this function-concept is clearly far less restricted.[44] We can read $\vdash\Phi(A)$ as '*A* has the property Φ', and $\vdash\Psi(A, B)$ as '*B* stands in the Ψ-relation to *A*' or '*B* is a result of an application of the process Ψ to the object *A*'.[45]

— 8[46] —

In the expression of a judgement, we can always regard the combination of signs which stands on the right of the sign of assertion as a function of one of the signs occurring therein. 'if we put at the place of this argument a German letter and in the content-line make a hollow in which the same letter stands, thus:

$$\vdash\!\!-\mathfrak{a}\!\!-\!\Phi(\mathfrak{a}),$$

this denotes the judgement that that function is a fact (*Thatsache*) whatever we may regard as its argument.'[47] It is to be noticed that the implication, which is easily expressed, 'the case of the affirmation of *X* ($\mathfrak{a}$) whatever $\mathfrak{a}$ is and the denial of *A* does not occur' does not exclude the possibility that $X(\varDelta)$ is affirmed and *A* denied. Here we cannot put anything in the place of $\mathfrak{a}$ without putting the correctness of the judgement in danger. This explains why the hollowing with the German letter within the hollowing is necessary:[48] it limits the domain to which the generality denoted by the letters refers. Only within its domain does the German letter keep its denotation;[49] in a judgement the same German letter can occur in different domains without the denotations assigned to it in one domain extending to the others. The domain of one German letter can include that of another, as the example '($\mathfrak{a}$)[($\mathfrak{e}$)$B(\mathfrak{a}, \mathfrak{e})$ implies *A* ($\mathfrak{a}$))'[50] shows. In this case, they must be chosen differently; we may never replace $\mathfrak{e}$ by $\mathfrak{a}$. Of course, we may replace a German letter, wherever it occurs in its domain, by another, if only at the places where different letters stood before, different ones stand afterwards. But other substitutions are only permissible if the hollowing immediately follows the sign of assertion, so that the domain of the German letter is the content of the whole judgement; and Frege[51] introduced the

---

43 ibid., pp. 17–18.

44 ibid., pp. 18, 19.

45 ibid., p. 18.

46 This article describes Frege's section on 'Die Allgemeinheit' (ibid., pp. 19–24). Cf. *F.u.B.*, pp. 23–7; *Gg.*, i, pp. 11–14; Russell, *Principles*, p. 519; and Schröder's comments in article 11 below.

47 *Bs.*, p. 19.

48 Without the hollowing at the right place, one would be in doubt as to whether '$\neg\Phi(x) = \Psi(x)$' was to be regarded as the denial of a generality or the generality of a denial; the domain of generality would not be sufficiently bounded (*Gg.*, i, p. 11).

49 'We cannot properly speak of the *denotation* of a German letter' (F., 1910).

50 For convenience in printing, I here, following Russell, replace the horizontal line containing '$\mathfrak{a}$' in a hollow by '($\mathfrak{a}$)'.

51 *Bs.*, p. 21. Cf. *Gg.*, i, pp. 31–4. A Latin letter can thus express the generality of a denial, but not the denial of a generality. See also Russell, *A.J.M.*, xxviii, pp. 192–202; xxx, pp. 231–6.

abbreviation of using a Latin letter, which meant that its domain was the content of the whole judgement. Thus, we can replace

$$\text{‘}{\vdash}X(a)\text{’ by ‘}{\vdash}(a)X(a)\text{’,}$$

if a only occurs at the argument-places in $X(a)$; and that, from ‘$\vdash$$A$ implies $\varPhi(a)$’ we can deduce ‘$\vdash$$A$ implies $(a)\varPhi(a)$’, if A is an expression in which a does not occur, and if a only stands in the argument-places in $\varPhi(a)$.[52]

The occurrence of the sign of negation to the left or to the right or on both sides of the hollowing, and of the line of condition to the right of the hollowing, were then treated.[53]

— 9 —

Frege[54] gave a representation and deduction, by means of the signs and rules introduced, of some judgements of pure thought. The derivation of the more complicated of these judgements from simple ones is not to make them more certain – for the most part, this would be unnecessary – but to allow the relations of the judgements to one another to appear;[55] and in this way we ultimately get a residuum of indemonstrable propositions which, when added to the laws contained in the rules for the use of the signs, include the content of all the laws of thought.[56] There are, perhaps, other sets of judgements which might form such a residuum,[57] Frege's own indemonstrables were nine in number,[58] and may be expressed by the formulæ:[59]

---

52 Peano (*R.d.M.*, v, pp. 126–7; cf. Frege, *R.d.M.*, vi, p. 59; see below) misunderstood Frege's use of German, Latin and Greek letters; imagining that they were implicit hypothesis. Frege's conventions here are not necessary, and were abandoned by Russell (*A.J.M.*, xxviii and xxx). Frege remarked (F., 1910) that 'it is correct that one can give up the distinguishing uses of Latin, German, and perhaps also of Greek letters, but at the cost of perspicuity of formulæ.'
53 *Bs.*, pp. 22–4. When the sign of negation is on the left of the hollowing in the assertion of $X(a)$ whatever a is, denotes the particular proposition: 'There are some things which do not have the property X.' If there is a sign of negation on each side of the hollow, the proposition: 'There is at least one X' is expressed. If to the right of the hollow the sign for what we have called '$X(a)$ implies $P(a)$' stands, the proposition may be read: 'every X is a P.'
54 *Bs.*, pp. 25–54. Cf. article 11 below.
55 *Bs.*, p. 25; cf. *Gl.*, p. 2.
56 Frege here used the phrase 'Gesetze des Denkens' (cf. *Gg.*, i, p. xv).
57 *Bs.*, p. 25. Cf. *BP*.
58 *Bs.*, p. 26.
59 ibid., pp. 26, 35, 43, 44, 47, 50, 51. In Boole's symbols we express these respectively as:

$$a\overline{b}\overline{a} = 0, \quad \overline{(c\overline{b}\overline{a})(c\overline{b}.c\overline{a})} = 0,$$

$$(d\overline{b}\overline{a})\overline{(bd\overline{a})} = 0, \quad (\overline{b}a)(a\overline{b}) = 0, \quad \overline{\overline{a}}a = 0, \quad a\overline{\overline{a}} = 0,$$

$$\overline{(c \equiv d)f(c)f(\overline{d})} = 0, \quad c \equiv c.$$

This last proposition cannot be expressed in Boole's symbols alone (cf. Schröder's review of Frege, referred to in article 11 below), and all Frege's propositions were stated with the sign of assertion, which Boole lacked.

(1) *a* implies that (*b* implies *a*);

(2) (*c* implies that (*b* implies *a*)) implies that ((*c* implies *b*) implies that (*c* implies *a*));

(3) (*d* implies that (*b* implies *a*) implies that (*b* implies that (*d* implies *a*));

(28) (*b* implies *a*) implies that (not-*a* implies not-*b*);

(31) not-not-*a* implies *a*;

(41) *a* implies not-not-*a*;

(52) $c \equiv d$ implies that $(f(c) \text{ implies } f(d))$;

(54) $c \equiv c$;

(58) $f(\mathfrak{a})$ holds whatever $\mathfrak{a}$ is, implies that $f(c)$ holds.

Finally,[60] the use of Frege's ideography in developing a general theory of series was shown.

A definition of an expression containing signs not explained before[61] is not a judgement.[62] A definition does not say: 'The right side of the equation has the same content as the left side', but 'it *is* to have the same content', and its purpose is merely abbreviation. A definition is, then, not an asserted proposition, and Frege put before a definition the double mark $\Vdash$ to indicate that a definition plays a double part in that it is at once transformed into a judgement when the expression defined is interpreted.[63]

— 10 —

Frege gave some examples of the expression of arithmetical and geometrical propositions by means of his ideography before the Jena scientific society in January 1879,[64] and, in January 1882, delivered a lecture on the aims of this ideography.[65]

— 11 —

Schröder[66] reviewed the *Begriffsschrift* soon after its publication. He remarked that Frege's ideography has almost nothing in common with Boole's[67] calculus of

---

60 *Bs.*, pp. 55–87, viii, i, Russell, *Principles*, p. 520.

61 *Bs.*, pp. 55–6. In this definition, Frege used small Greek letters (*Bs.*, pp. 56–61).

62 Frege had a reference to the Kantian opinion that all the judgements of arithmetic are synthetic on p. 56 of the *Bs*.

63 Cf. *Gg.*, i, pp. 43–52; ii, pp. 69–80, for a fuller treatment of definitions and the rules to be observed in making them. Cf. also the section on Peano below.

64 'Anwendungen der Begriffsschrift', *Sitzungsber. der Jenaischen Ges. für Medicin und Naturwiss.*, 10 January 1879, 5 pages.

65 'Über den Zweck der Begriffsschrift', 27 January ibid., 1882, pp. 1–10.

66 *Zeitschr. für Math. und Phys.* (*Historisch-literarische Abtheilung*), xxv, 1880, pp. 81–94.

67 Schröder remarked that the work of others, especially of Boole, was unnoticed by Frege, except perhaps in the passage referred to above on p. iv of the introduction to the *Begriffsschrift*.

concepts, but it has with Boole's calculus of judgements; and Schröder gave [68] an account of how Boole founded the latter on the former by considering the times during which the judgements[69] *a, b, c. ...* are true.[70] Then, where 1 stands for the

---

68 loc. cit., p. 87.

69 Instead of 'judgements' it would be better here to say 'thoughts'. That of which one asserts (*aussagt*) the truth is not a judgement, but a judicable content, a thought. A thought is not true at one time and false at another, but is either true or false – *tertium non datur*. The false appearance that a thought can be true at one time and false at another arises from an incomplete expression. A complete proposition (*Satz*), or expression of a thought (*Gedankenausdruck*), must also contain the time-datum. If we say: 'The Elbe has risen one metre above the zero of the gauge at Magdeburg,' the time belongs to the thought-content of the proposition if what is said is the case. But the truth is timeless.

More correctly, what is meant would be thus expressed. Let the function $\Phi(\xi)$ have the truth as value for some moments of time as arguments, and the false for others, and the false for all arguments which are not moments of time. Then we have a class of moments of time $(\grave{\varepsilon}\Phi(\varepsilon))$, for which the value of the function is the truth.

But the intermixture of the time appears to me not suitable (*sachgemäss*). And now we know that, when classes are introduced, a difficulty (Russell's contradiction) arises. In my fashion of regarding concepts as functions, we can treat the principal parts of Logic without speaking of classes, as I have done in my *Begriffsschrift*, and that difficulty does not then come into consideration. Only with difficulty did I resolve to introduce classes (or extents of concepts), because the matter did not appear to me quite secure – and rightly so, as it turned out. The laws of numbers are to be developed in a purely logical manner. But numbers are objects, and in logic we have only two objects, in the first place: the two truth-values. Our first aim, then, was to obtain objects out of concepts, namely, extents of concepts or classes. By this I was constrained to overcome my resistance and to admit the passage from concepts to their extents. And, after I had made this resolution, I made a more extended use of classes than was necessary, because by that many simplifications could be reached. I confess that, by acting thus, I fell into the error of letting go too easily my initial doubts, in reliance on the fact that extents of concepts have for a long time been spoken of in Logic. The difficulties which are bound up with the use of classes vanish if we only deal with objects, concepts and relations, and this is possible in the fundamental part of Logic. The class, namely, is something derived, whereas in the concept – as I understand the word – we have something primitive. Accordingly, also the laws of classes are less primitive than those of concepts, and it is not suitable to found Logic on the laws of classes. The primitive laws of Logic may contain nothing derived. We can, perhaps, regard Arithmetic as a further-developed Logic. But, in that, we say that in comparison with the fundamental Logic, it is something derived. On this account I cannot think that the use of arithmetical signs ('+', '−' ':') is suitable in Logic. The sign of equality is an exception; in Arithmetic it denotes, at bottom, identity, and this relation is not peculiar to Arithmetic. It must be doubtful *a priori* that it is suitable to constrain Logic in forms which originally belong to another science. (F., 1910.)

70 On p. 87, Schröder mentioned MacColl's researches, and on p. 94 found fault with MacColl's use of symmetrical signs (: and ÷) to represent the unsymmetrical relation of subordination.

whole period of time considered in the presuppositions of the investigation,[71] he expressed many of Frege's propositions in the Boole-Schröder form.[72]

Thus, a, preceded by Frege's sign of assertion, purported that a holds; and Boole represented this by $a = 1$ or $\bar{a} = 0$.[73] The relation 'if b holds, then a holds;[74] of Frege was represented by $\bar{a}b = 0$ or $a + \bar{b} = 1$; and similarly Frege's schemes for representing the assertions 'b and c together imply a' and 'c implies b, implies a' became respectively $\bar{a}bc = 0$ and $\bar{a}(\bar{b}c) = 0$.[75] The most important propositions in the second section of Frege's work were then expressed in the markedly shorter form in which the Boole-Schröder notation enables them to be, and Schröder[76] pointed out that Frege made frequent repetitions, which a scheme showing commutativity or the property of double negation would avoid, and used[77] superfluous premises – premises which are contained in others.

---

71 These suppositions of the investigation are properly conditions which, when made explicit as conditional propositions, are to be added to all propositions which occur in the investigation. To separate out some conditions and to put them down as suppositions of the investigation is quite arbitrary. If we start from other suppositions, we change thereby the denotation of the sign '1'. I do not recognize such a separation of the suppositions of the investigation from the other conditions, and for this reason my propositions do not wholly coincide with those with which Schröder wished to replace them. The use of the sign '1' in the stated manner is only possible in a logic which is kept strictly separated from Arithmetic, and with which, therefore, an application in Arithmetic is excluded, because otherwise the uniqueness of the sign, which, besides this, was already endangered, would be lost.

   To found the 'calculus of judgements' on the 'calculus of concepts' (which is properly a 'calculus of classes') is to reverse the correct order of things; for classes are something derived, and can only be obtained from concepts (in my sense). But concepts are something primitive which cannot be dispensed with in Logic. We can only determine a class by giving the properties which an object must have in order to belong to the class. But these properties are the attributes (*Merkmale*) of a concept. We define a concept, and pass over from it to the class. For that reason, calculation with classes must be founded on the calculation with concepts. And the calculation with concepts is itself founded on the calculation with truth-values (which is better than saying 'calculus of judgements'). (F., 1910.)

72 loc. cit., pp. 87–91.

73 If my '$\vdash a$' is reproduced in the form '$a = 1$', the assertory force lies in the sign of equality. But then '$a = 1$' would never be capable of being used without assertory force, as a proposition expressing a condition or a consequence, for example. F., 1910.

74 'If, indeed, not necessarily, at least de facto (*factisch*)', Schröder, ibid., p. 88.

75 On pp. 88–9, Schröder pointed out that Frege's (op. cit., p. 7) first interpretation of the second implication as $\bar{a}\bar{b}c = 0$ is wrong, but his second $\overline{\bar{a}b}c = 0$ is consistent with the applications made of it.

76 loc. cit., p. 91.

77 Frege, op. cit., Nos. 3, 4, 32, 45.

Dealing next with Frege's section on generality, Schröder[78] showed how modifications or extensions can easily be made in Boole's symbolism to make it express particular judgements. If '$f(a) = 1$' means 'all a's have the property f', then '$\{f(a)\}_1 = 1$', or more shortly '$f_1(a) = 1$', will mean 'all a's have the property not-f', and '$\overline{f(a)} = 1$' will mean 'not all a's have the property f'. If '$P(a) M(a) = 0$' means 'no M is a P', '$P(\bar{a}) M(\bar{a}) = 0$' will deny that 'no M is a P' for *every* a.

On Frege's section 'Die Function', Schröder[79] remarked: 'The interpretation the author gives to the concept of a (logical) "function" is quite peculiar and very comprehensive. It goes much further than all previous explanations and does not seem to me to be without justification. However, because of limitations of space, I will refer in this connection to the book itself ...' Referring to Frege's Appendix on the theory of series, Schröder[80] merely expressed a wish for a simpler symbolism.

Thus Schröder[81] concluded that, with the exception of the sections on the function, on generality, and on the theory of series, Frege's work was devoted to the foundation of a symbolic system (*Formelsprache*) which coincided in essentials with Boole's manner of representation of, and calculation with, judgements, and which did not advance beyond it.

— 12 —

This seems to be the most suitable place for giving some account of Peano's views of Frege's symbolism, and Frege's account of the difference in aim between his own and other systems of ideography. The motive of these discussions was afforded by a review of the first part of Frege's *Grundgesetze der Arithmetik* of 1893, which was written by Peano.[82] Although the doctrines contained in this later work of Frege's

---

78 loc. cit., pp. 91–2. Cf. the section on Boole, above.

On Schröder's proposals, I make the following remarks. The asserting force, which with my notation is in the line of judgement, must here be found in the sign of equality. Hence it follows that theorems like '$f(a) = 1$', '$f_1(a) = 1$' may never occur without being asserted, and therefore never as conditional theorems or deductions, and this is a great defect. The limitation of the domain of the German letters is not clearly to be apprehended. We have here three notations for negation: the '0', the index '1' in 'f_1', and the line over the 'a'. By this the matter becomes very complicated. Instead of having to do with one sign, we must now establish rules for three signs. This superfluity of signs is a disadvantage. As long as one has in view merely a representation of the simplest logical laws, Schröder's proposals may appear not useless; but they would fail entirely if one wished to represent more complicated logical structures – actually to carry out mathematical proofs with them. In comparison with Boole's Symbolic Logic my ideography appears cumbrous, if one considers it merely as Symbolic Logic – as a *calculus ratiocinator* and not as a *lingua characterica*. But this disadvantage becomes an advantage if one uses it for its proper purpose. It is precisely this cumbrousness that makes it possible for the eye to take in in one glance – at least as regards the principal features – a complex logical structure. By means of this cumbrousness, the more complicated formulæ gain a perspicuity that would not be reached without it, and then often would a chaos present itself to the eye, which could hardly be extricated from confusion (F., 1910).

79 loc. cit., p. 92.
80 ibid., pp. 92–3.
81 ibid., pp. 83–4.
82 *R.d.M.*, v, 1895, pp. 122–8.

are a great advance on the doctrines of the *Begriffsschrift* and the *Grundlagen der Arithmetik* of 1884,[83] Peano's remarks almost wholly referred to those parts of the symbolism and those conceptions of the *Grundgesetze* which are possessed also by the *Begriffsschrift*.

Peano compared Frege's notations with his own,[84] and concluded that Frege's system was inferior to his, both from the scientific and from the practical point of view. Practically, it was rightly urged by Peano[85] that, in Frege's representation of logical multiplication, sight was lost of its commutative and associative properties; but Peano's[86] claim that his system corresponds to a more profound analysis, because it only uses three fundamental signs whereas Frege's system is based on five, cannot be substantiated, because, firstly, Frege's notation expressed some ideas which Peano's was unable to do, and secondly, Peano's enumeration of the fundamental signs he used was incomplete.[87]

According to Peano,[88] where '*a*' represents a proposition, Frege's notation of '*a*' preceded by the sign of assertion denotes '*a* is true', and '*a*' preceded by a horizontal line alone denotes 'the truth of *a*'. Peano could not see the utility of these conventions, which had nothing to correspond to them in the *Formulaire*; thus '*a*' alone there signified '*a* is true', and, in implications, the truth neither of the hypothesis nor the thesis was asserted, but only that of the main implication. Frege's sign for the unasserted 'a_x whatever x is',[89] where 'a_x' denotes 'a proposition containing a variable letter x', was interpreted by Peano in his own symbols as

$$V = {}_x a_x,$$

and Frege's implications between propositional functions were interpreted in an analogous way.[90]

Finally, Peano[91] remarked that 'certain distinctions are difficult to grasp, since often two German terms, between which Frege distinguished, correspond in the dictionary to the same Italian term.' These words indicated the series of subtle distinctions whose introduction into logic formed a great part of Frege's work subsequent to the *Begriffsschrift*: – The distinction of 'Sinn' and 'Bedeutung', of the three elements in judgement,[92] and 'Begriff' and 'Gegenstand'. Before, however, attempting an account of these, we will consider, firstly, Frege's[93] own account of the aims and pecularities of his ideography, and, secondly, his theory of arithmetic as given in the *Grundlagen der Arithmetik* of 1884.

----

83 This work will be described below.
84 Cf. the section on Peano, below.
85 loc. cit., p. 125.
86 ibid.
87 Cf. Frege, *R.d.M.*, vi, 1896, pp. 53–9. This reply to Peano's criticism will be referred to further below. See also the section on Peano, below.
88 loc. cit., p. 124.
89 The use of Latin letters here corresponds to Frege's use of German ones.
90 Peano, loc. cit., pp. 124–5.
91 ibid., p. 127.
92 *Gg.*, i, p. x.
93 'Ueber die Begriffsschrift des Herrn Peano und meine eigene', *Ber. der math.-phys. Cl. der Kgl. Sächs. Ges. der Wiss. zu Leipzig*, Sitzung vom 6. Juli 1896, pp. 361–78.

After an account[94] of the necessity of an ideography, a *lingua characteristica*, as Leibniz expressed it, in the question as to the completeness of a list of axioms in a certain branch of knowledge, Frege explained[95] his aim: 'The object of my intention is therefore absolute rigour in demonstration and the greatest logical precision, and besides, perspicuity and brevity.' Peano's ideography was a descendant of Boole's logical calculus, but, whereas Boole's logic was logic and nothing else, Peano's object was to pour a content into this form; and, in this respect, his undertaking stood nearer to that of Frege than the logic of Boole:[96] Boole's logic is a *calculus ratiocinator* but not a *lingua characterica*; Peano's mathematical logic is principally a *lingua characterica* and incidentally also a *calculus ratiocinator*, while my conceptual notation is meant to be both with equal emphasis.'

Frege[97] also remarked that the two-dimensional character of his notation gives it the advantage, over the notations in which all is written in a line,[98] of making it possible to unify, by joining the lines denoting logical relations, the different parts of a theorem. It is true that the single-line systems are more convenient to print, but 'the convenience of the type-setter is surely not the greatest of goods.' The more specific criticism of Frege on Peano's system, and Peano's answers to these criticisms, will be dealt with when we shall have considered fully Peano's system.

— 13 —

In 1884 Frege published, under the title of *Grundlagen der Arithmetik*,[99] a critical account of the views of many philosophers and mathematicians on various arithmetical topics, and a short but clear statement in ordinary language of his logical theory of natural numbers and the other 'numbers' of arithmetical analysis.

The striving, said Frege,[100] in modern mathematics, after rigour of proofs, accurate determination of limits of validity, and, for this purpose, clearness of concepts, must, at last, extend to the concept of number in the sense of 'number given by enumeration' (*Anzahl*)[101] and the simplest laws which hold for positive integers. The object of a proof is not merely to raise the truth of a theorem above every doubt, but also to impart an insight into the dependence of truths on one another. The further one continues these investigations, the fewer the fundamental truths to which one can reduce everything; and this simplification is in itself an end worthy to be striven for.[102]

---

94 ibid., pp. 362–5.
95 ibid., p. 365.
96 ibid., p. 370. Cf. *Gl.*, p. 103n.
97 *B.P.*, pp. 364, 378.
98 Such notations were Boole's, MacColl's, and, above all, Peano's.
99 *Die Grundlagen der Arithmetik. Eine logisch mathematische Untersuchung über den Begriff der Zahl* (Breslau 1884), xi, 119 pages.
100 ibid., pp. 1–2.
101 This is usually called *cardinal number*, but as, chiefly since Cantor's paper of 1895 (*Math. Ann.*, xlvi), this term has obtained a meaning which is independent of a process of enumeration, we shall translate *Anzahl* by *enumeral*.
102 ibid., p. 2.

It was this desire for simplification, together with the philosophical questions as to the *a priori* or *a posteriori*, synthetic or analytic,[103] nature of arithmetical truths, which moved Frege[104] to his investigation. If, in our proofs of mathematical truths, we only encounter the laws of logic and definitions, we have an analytic truth; but if it is not possible to carry out the proof without using principles which are not, in general, of a logical nature, but refer to a special department of knowledge, the theorem is synthetic.

— 14 —

Already, in the first words of the introduction, Frege,[105] by pointing out that the statement 'the number 1 is a thing' is no definition, because on the one side of the copula stands the definite article and the other side the indefinite article, so that it does not state *which* thing the number 1 is, showed a degree of subtlety not attained even now by some mathematicians who have not paid particular attention to mathematical logic; and Frege's acuteness was further shown in his criticism of the psychological standpoint in mathematics,[106] and of the views of many philosophers and mathematicians on various parts of the theory of arithmetic.

In this critical part, Frege dealt with the views (in especial) of Kant, Hankel, Leibniz, H. Grassmann and J. S. Mill on the provability of formulæ about particular numbers, like $2 + 3 = 5$;[107] the views of Mill and Leibniz as to whether the laws of arithmetic are inductive truths;[108] the views of Kant, Baumann, Lipschitz, Hankel, Leibniz, Jevons and Mill as to whether these laws are synthetic *a priori* or analytic;[109] the views of Newton, Hankel, Leibniz, M. Cantor, Schröder,[110]

---

103 These (Kantian) distinctions are not concerned with the content of the judgement but with the justification of a passing of judgement, since, where this is lacking, the possibility of the distinction referred to vanishes. An error *a priori* is then such a non-entity (*Unding*) as a blue concept. If one calls a theorem *a posteriori* or analytic, one does not judge about the psychological, physiological and physical conditions which have made it possible to form the content of the theorem in consciousness, nor about how another person has – perhaps in an erroneous manner – arrived at maintaining its truth, but on what, at bottom, the justification of the maintenance of its truth rests. A truth is *a posteriori* if its proof must depend on facts, that is to say, unprovable truths without generality which contain statements about definite objects. If, on the other hand, it is possible to carry out the proof wholly from general laws which themselves neither are capable of proof nor need it, the truth is *a priori* (ibid., pp. 3–4). This is not quite Kant's distinction, but an improvement upon Kant's more limited view that the sole source of analytic judgements is the principle of contradiction (cf. Couturat, *Les Principes des Mathématiques. [Avec une appendice sur la philosophie des mathématiques de Kant]* (Paris 1905 [repr. Hildesheim 1965], pp. 235–308).

104 ibid., pp. 3–5. Cf. article 1 above.

105 op. cit., p. 1.

106 *Gl.*, pp. v–x, 105; cf. *Gg.*, i., pp. xiv–xxvi.

107 *Gl.*, pp. 5–12.

108 ibid., pp. 12–17.

109 ibid., pp. 17–24.

110 Cf. Schröder's note in his *Vorlesungen über die Algebra der Logik*, i (Leipzig 1890), p. 704.

Baumann, Mill, Locke, Berkeley, Lipschitz, Schlömilch and Thomae on the concept of enumeral (a general concept, as opposed to those of the individual numbers such as 3 and 4);[111] and the views of Schröder, Leibniz, Baumann, Locke, G. Köpp, Hobbes, Hume, Thomae, Descartes, Jevons, Hesse, Hankel and Spinoza on questions relating to the Unit and the One (*Einheit und Eins*).[112]

— 15 —

In treating the concept of enumeral,[113] Frege's starting-point was that the number-datum contains a statement about a concept.[114] He first attempted to complete Leibniz's[115] definitions of the single numbers by definitions of 0 and 1; thus he defined:[116] 'a concept has the number 0' to mean 'whatever a is, a does not fall under this concept', and 'a concept F has the number 1' to mean 'it is not generally true that, whatever a is, a does not fall under F' together with 'from the theorems "a falls under F" and "b falls under F", it quite generally follows that a and b are the same'. Further, the passage from a number to the next one was thus universally defined: A concept F has the number $(n + 1)$ if there is an object a which falls under F and is such the concept 'falling under F but not a' has the number n.

But firstly, though, by these definitions, we can say what is the meaning of 'F has the number $1 + 1$', 'F has the number $1 + 1 + 1$', and so on, we cannot decide if, for example, 'a concept has the number Julius Caesar', that is to say, whether Caesar is a number or not.[117] Secondly, we can never prove with these definitions that, if F has the number a and also has the number b, then a is equal to b. Thus the expression '*the* number of F' is not justified: we have only defined such phrases as 'has the number 0 (or 1)', and it is not allowable to distinguish the 0 or the 1 as independent objects which can again be recognized.[118]

As numbers are independent objects, we must have a criterion for recognizing them (in a logical sense) when we meet them again.[119] Now, such a means – that is to say, a means of determining the equality or not of the numbers of the concepts F

---

111 Frege, *Gl.*, pp. 24–39. Frege concluded (cf. ibid., pp. 58, 115ff that an enumeral is neither a heap (*Haufe*) of things (ibid., pp. 38–9), nor a property of such (ibid., pp. 27–33), nor is it something subjective (ibid., pp. 33–8), but the *datum* of the number of a concept states something objective.

112 Whether the word 'one' expresses a property of objects (ibid., pp. 39–44); whether units are equal to one another (ibid., pp. 44–51); attempts to get over the difficulty of reconciling the equality and distinguishability of units (ibid., pp. 51–8); and solution of the difficulty (ibid., pp. 58–67, especially pp. 66–7).

113 ibid., pp. 67–99.

114 ibid., pp. 67, 68–72.

115 ibid., p. 7. These definitions were $2 = 1 + 1, 3 = 2 + 1, \ldots$.

116 ibid., p. 67.

117 The apparent triviality of this statement only disappears on reflection. In fact, the numbers are not, by the above definitions, defined as entities which can be distinguished from all other entities.

118 ibid., p. 68.

119 ibid., p. 73.

and G – is known and is what mathematicians call a (1, 1) correspondence.[120] After discussing equality and its re-definition for the various clauses of entities dealt with in mathematics,[121] Frege defined the enumeral of the concept as the extension of the concept 'having a (1, 1) correspondence with F (*gleichzahlig*[122] *mit F*)'.[123]

The notions of *relation* and correspondence by means of a relation belong to logic.[124]

Definitions of 0,[125] of the expression 'n follows m immediately in the number-series',[126] of 1,[127] of succession in a series,[128] of finite number,[129] and of infinite enumerals[130] followed.

— 16 —

Finally, Frege dealt with Kant's underestimation of the value of analytic judgements,[131] the necessity for an ideography,[132] the definitions of other (fractionary, irrational and complex) numbers,[133] and concluded with a *résumé*.[134]

―――――――

120 ibid., pp. 73–4.

121 *Gl.*, pp. 74–9. Cf. the discussions of Frege and Peano in *R.d.M.*, vi, No. 2 (1898), pp. 53–61 [= *Lettera del sig. G. Frege all' Editore* (pp. 53–9) and *Risposta Peanos* (pp. 60–1)]; in *B.P.*; and in *Gg.*, ii, pp. 70ff: the section on Peano, below; and a paper by Burali-Forti (see the section on Peano's school, below).

On the principle of permanence, see Frege, *Ueber die Zahlen des Herrn H. Schubert*, (Jena 1899); Russell, *Principles*, p. 377; and Couturat, *Les Principes*, pp. 89–90.

122 The *word*, but *merely* the word, implies a circle in the definition of 'Zahl' (cf. ibid., [= *Couturat, Les Principes*] p. 26; and Schröder, op. cit., i, p. 704). The notion of a (1, 1) correspondnce does not presuppose the notion of the number 1 (see Russell, *Principles*, p. 113: Couturat, *Les Principes*, p. 32). Cantor used the word 'äquivalent', Dedekind and Russell used 'ähnlich' and 'similar' respectively.

123 *Gl.*, pp. 79–81.

124 ibid., pp. 81–6. Cf. *Gg.*, i, p. 3.

125 *Gl.*, pp. 86–9.

126 ibid., pp. 89–90.

127 ibid., pp. 90–2.

128 ibid., pp. 92–5.

129 ibid., pp. 95–6.

130 ibid., pp. 96–9. This is Cantor's first transfinite 'power' (*Mächtigkeit*) or 'cardinal number' (ibid., p. 108n; *Gg.*, i, p. v).

On Frege's theory of progressions and arithmetic, see Russell, *Principles*, p. 520.

131 On Kantianism in mathematics, see also Couturat, *Les Principes*, pp. 235–308.

132 *Gl.*, pp. 99–104.

133 ibid., pp. 104–15.

In his treatment of the questions of existence of mathematical objects and compatibility (*Gl.*, pp. 65, 105–8, 87), Frege's views coincided with those later expressed by Mario Pieri (*R.M.M.*, March 1906, p. 196 [*Sur la compatibilité des axiomes de l'arithmétique*, l.c., pp. 196–207]).

On pp. 114–15 of the *Gl.*, Frege briefly indicated that fractions, complex numbers, etc., were to be defined as classes.

134 *Gl.*, pp. 115–19.

— 17 —

In the *Grundlagen*, Frege had criticized severely the formalist theory of arith-metic[135] according to which numbers are empty signs to which certain rules of operation are assigned much as the moves to chess pieces. In 1885[136] he dealt with the question again, and contrasted it with the view (also called 'formal'), which he held, that all arithmetical theorems can be derived in a purely logical manner from definitions. Frege often returned to the question of the confusion of symbols with the objects of mathematics; namely, in an article of 1895,[137] in the ironical pamphlet[138] on H. Schubert's exposition[139] of the foundations of arithmetic, and in the criticism[140] of the theories of irrational numbers of Heine, Cantor and Thomae.

— 18 —

In 1891, Frege delivered a lecture entitled *Function und Begriff*,[141] in which his developed views on functions and generality were given.

The definition of a function of x, said Frege, as an expression of calculation which contains x, is not satisfactory because there sign and what is signified are not distinguished.[142] But if the function were only what the expression of calculation

---

135 *Gl.*, pp. 107–08.
136 'Ueber formale Theorien der Arithmetik', *Sitzungsber. der Jenaischen Ges. für Medicin und Naturwiss.*, 17 July 1885, 11 pages.
On p. 4, Frege mentioned the view that real numbers are signs, while integers are not. This view seems to have been actually held by Georg Cantor.
137 'Le nombre entier', *R.M.M*, January 1895, pp. 73–8. This article also contains other criticisms on articles in the same periodical by Ballue, and by Le Roy and Vincent.
L Couturat, on pp. 89–91 of his article 'Contre le nominalisme de M. Le Roy' (*R.M.M.*, viii, 1900, pp. 87–93), dealt with the confusion of *symbols* with numbers. Cf. pp. 225, 226–8 of E. Le Roy's 'Réponse à M. Couturat' (ibid., pp. 223–33).
138 *Ueber die Zahlen des Herrn H. Schubert* (Jena 1899), vi, 32 pages. A short account of the contents of this pamphlet was given in the *R.M.M.*, March 1900, Supplément, pp. 4–5.
139 'Grundlagen der Arithmetik', *Encykl. der math. Wiss.*, I. A 1, [Leipzig] 1898 [pp. 1–27].
140 *Gg.*, ii, 1903, pp. 72–4, 80–139; cf. ibid., i, 1893, p. xiii. To this refer the articles of Thomae, 'Gedankenlose Denker. Eine Ferienplauderei', *Jahresber. der Deutsch. Math.-Ver.*, xv, 1906, pp. 434–8; Frege, 'Antwort auf die Ferienplauderei des Herrn Thomae', ibid., pp. 586–90; Thomae, 'Erklärung', ibid., pp. 590–2; Frege, 'die Unmöglichkeit der Thomae'schen formalen Arithmetik aufs Neue nachgewiesen', ibid., xvii, 1908, pp. 52–5; Thomae, 'Bemerkung zum Aufsatze des Herrn Frege', ibid., p. 56; and Frege, 'Schlussbemerkung', ibid.
141 *Function und Begriff. Vortrag gehalten in der Sitzung vom 9. Januar 1891 der Jenaischen Gesellschaft für Medicin und Naturwissenschaft* (Jena 1891), i, 31 pages (published separately).
142 Frege (ibid., p. 3) referred to the essays of Helmholtz and Kronecker on the number-concept (1887) as shown how 'the present very widespread tendency to recognize nothing as an object which cannot be perceived by the senses leads to the opinion that the number-signs are the numbers, the real objects of consideration.' Cf. *Gg.*, i, p. xiii.
That definitions are not *creative* in the sense that they give properties to things which have not got them, Frege strongly maintained (*Gl.*, pp. 107–8; *F.u.B.*, p. 4; *Gg.*, i, pp. xiii–xiv; ii, pp. 140–9; Russell, *Principles*, p. 451).

denoted, it would be a number, and nothing would be gained by it for arithmetic.[143] The essence of the function lies in that part of the expression which does not contain the x; and the argument does not belong to the function, but forms, together with the function, a complete whole: functions alone are incomplete (*ungesättigt*), and thereby differ fundamentally from numbers.[144] It is not correct to say that the function $2 + x - x$, for example, is 2. Although the value of the function is always 2, the function itself must be distinguished from 2, for the expression of a function must always exhibit one or many places to be filled up by the sign of the argument.[145]

'If we write $x^2 - 4x = x(x - 4)$, we have not put one function equal to the other, but only the function-values equal to one another; and if we so understand this equation that it is to hold whatever is put for x, we have thereby expressed the generality of an equation. But we can always say for it: "the range of the function $x(x - 4)$ is equal to that of the function $x^2 - 4x$", and thus have an equation between ranges. Now that it is possible to view the generality of an equation between function-values as an equation between ranges cannot, it seems to me, be proved, but must be regarded as a fundamental law of logic.'[146] Frege's[147] notation for the range of a function was got by replacing the sign of the argument by a Greek vowel, putting the whole in brackets, and putting before the bracket that vowel surmounted by a smooth breathing (*spiritus lenis*); thus the above equation may be written

$$\acute{\varepsilon}(\varepsilon^2 - 4\varepsilon) = \acute{\alpha}(\alpha(\alpha - 4)).$$

When mathematical modes of thought and expressions were transferred to logic, it was almost inevitable that functional expressions should be introduced in logic as 'any expressions involving x'. We have seen above that such expressions were considered by Boole, MacColl and Frege; and Frege, in the lecture which we are now considering,[148] emphasized that this extension went further in both of the directions[149] in which the signification of the word 'function' has been widened:

---

143 *F.u.B.*, p. 5. If, as is usual when we use the word 'function,' we think of expressions in which a number is only indefinitely indicated by x, nothing essential is altered; the expression still indicates a number, only indefinitely (ibid., pp. 5–6).

144 ibid., p. 6; *Gg.*, i, p. 6.

145 *F.u.B.*, p. 8.

146 ibid., pp. 9–10; cf. *Gg.*, i, pp. vii, 36. This indemonstrable premise played a part later in Frege's discussion of Russell's contradiction (*Gg.*, ii, pp. 253–65; *R.Pr.*, p. 522).

I have translated, following Russell (*R.Pr.*, p. 511), *Werthverlauf* by *range*.

147 F.u.B., p. 10; *Gg.*, i, pp. ix–x. The smooth breathing may be read *such that*, and corresponds to Peano's ɔ. Note that the Greek vowels are what Peano called *apparent* (any one, or pair, ..., can be replaced by any other one, or pair, ..., without altering the value of the formula; just like the 'x' under the integral sign in a definite integral).

For a critical discussion of Frege's ranges and cardinal numbers, see *R.Pr.*, pp. 132, 510–18, 519.

148 *F.u.B.*, pp. 12–13, 17; cf. *Gg.*, i, p. 6.

149 Namely, in the process used or imagined for defining functions (cf. Dirichlet's definition), and in considering also complex values as arguments and function-values.

signs like '=', '>', '<' besides '+', '−', ..., were used to form a functional expression, so that '$x^2 = 1$' is a sign of a function; and not merely numbers, but any objects, can be arguments.

In the above function $x^2 = 1$, we may substitute various arguments (for example, −1, 0, 1, 2) for x, and get true or false equations. Now, Frege[150] used the phrase: 'the value of our function is a truth-value (*Wahrheitswerth*)', distinguished the truth-value of the true from that of the false, and considered that '$2^2 = 4$' denotes that of the true[151] just as '2^2' denotes[152] 4. Thus, we have a true equation in $(2^2 = 4) = (2 > 1)$, for both sides denote the same thing.

'Here the objection lies ready to hand that "$2^2 = 4$" and "$2 > 1$" express quite different thoughts; but also "$2^4 = 4^2$" and "$4.4 = 4^2$" express different thoughts, and yet we can replace "2^4" by "4.4" because both signs have the same signification. Consequently "$2^4 = 4^2$" and "$4.4 = 4^2$" have the same denotation. Hence we see that equality of thought does not follow from equality of denotation. If we say: "the evening star is a plant whose period of revolution is less than that of the earth," we have expressed another thought than that expressed in the sentence: "the morning star is a planet whose period of revolution is less than that of the earth"; for he who does not know that the morning star is the evening star could hold that one was true and the other false; and yet the denotation of both propositions must be the same, because only the words "morning star" and "evening star" are interchanged, and they have the same denotation, that is to say, they are proper names of the same celestial body. We must distinguish meaning (*Sinn*) and denotation (*Bedeutung*),[153] "2^4" and "4.4" have the same denotation, that is to say, they are proper names of the same number, but they have not the same meaning; and on this account "$2^4 = 4^2$" and "$4.4 = 4^2$" have the same denotation, but not the same meaning; that is to say, in this case they do not contain the same thought'.

---

150 *F.u.B.*, p. 13.

151 For the sake of shortness, Frege said '"$2^2 = 4$" denotes the true' and '"$2^2 = 1$" denotes the false."

152 Like Frege (cf. *Gg.*, i, p. 4), we here use inverted commas to denote that we are speaking of the *sign*, and we do not use them when we speak of its denotation.

153 Here and elsewhere I have translated *bedeuten* and *Bedeutung* by *denote* and *denotation* respectively; Russell (op. cit., p. 502) translated *Bedeutung* by *indication*.

Frege developed this distinction at length in an essay 'Über Sinn und Bedeutung' (*Zeitschr. für Philos. und phil. Kritik*, C. (1892), pp. 25–50; cf. *Gg.*, i, pp. x, 6–7). For a critical account of this distinction, and of Frege's analysis of judgement into (1) recognition of truth, (2) Gedanken, (3) truth-value, see *R.Pr.*, pp. 502–05.

Venn (op. cit., 1st ed., p. xxv, 2nd ed., p. xxvi) asserted that the distinction between denotation and connotation and the doctrine of definition are found to be of necessity passed by, by symbolic logic. That this statement is mistaken with regard to the doctine of definition, we learn from the work of Frege and Peano (see above, and the section on Peano below); that it is also mistaken with regard to the distinction between connotation and denotation, and the closely-allied distinctions between 'indication by form of the nature of operations' and denotation of signs (Venn, op. cit., 1st ed., pp. 32–3; 2nd ed., pp. 33–4) also appears. Cf. Miss E. E. C. Jones's distinction between a name's *signification* (or attribution) of some characteristics of that to which it applies and its *application* (or denomination) to

— 19 —

A concept (*Begriff*) in logic is a function whose value is always a truth-value,[154] and the extension of a concept is the range of a function whose value for every argument is a truth-value.[155]

For argument we have considered not merely numbers but all objects; while for function-values we have introduced the two truth-values. But[156] 'we must go further, and permit objects without limitation to be function-values.' For example, 'the capital town of *x*' was called by Frege[157] the expression of a function, since it is incomplete but is completed when the empty place marked by *x* is filled by a word denoting something complete, like 'the German empire'. When this particular argument is used, the function-value is Berlin. Thus objects without limitation were permitted both as arguments and as function-values, and, with Frege,[158] an *object* (*Gegenstand*), though a formal definition of what is meant is impossible because it cannot be decomposed logically, may be described as: everything which is not a function, and consequently whose expression has no empty place.

More accurate settlements about the denotations of the usual signs must now be made.[159] As long as whole numbers alone of objects are considered, the plus-sign need only be defined when it stands between whole numbers; but every extension of the class of objects considered necessitates a new definition of this sign. 'To make arrangements that an expression should never become devoid of signification, in order that we may never, without remarking it, calculate with empty signs while

---

some thing or group of things (*Elements of Logic as a Science of Propositions* (Edinburgh 1890), pp. 8, 8n, 196–7; *An Introduction to General Logic* (London 1892), p. 5; cf. J. N. Keynes, *Studies and Exercises in Formal Logic*, ..., 4th ed. (London 1906), pp. 43, 189–90), and Mrs Sophie Bryant's distinction between a symbol *representing* the operation of the mind on a certain subject-matter, and its *denoting* the result of that operation (*Proc. Aristot. Soc.*, ii, N.S. (1901–02) [*Sophie Bryant: The Relation of Mathematics to General Formal Logic*, pp. 105–34; and *Discussion on Mrs Bryant's Paper*, '*The Relation of Mathematics to General Logic*.' pp. 134–43], pp. 108, 139–40 [contribution by E. C. Benecke], 143).

In his *Bs.*, Frege supposed that identity is a relation between names of objects, but when he had distinguished *meaning* and *denotation*, so that, for example, 'the meeting point of the lines *a* and *b*' may have the same *denotation* as 'the meeting point of the lines *c* and *d* (which are different from *a* and *b*)', but a different *meaning* (be obtained by different constructions), he expressed his view on identity as follows (*S.u.B.*, p. 50): If *a* = *b*, the denotation (truth-value) of '*b*' is the same as that of '*a*', and thus the truth-value of '*a* = *b*' is the same as that of '*a* = *a*'. Nevertheless the meaning of '*b*' can differ from that of '*a*', and accordingly the thought expressed by '*a* = *b*' can be different from that expressed by '*a* = *a*'.

Of the same nature was Miss Jones's treatment of identical propositions (*Elements*, p. 46; *Gen. Logic*, pp. 20, 27–8). Cf. also *R.Pr.*, p. 64.

154 *F.u.B.*, p. 15.
155 ibid., p. 16.
156 ibid., p. 17.
157 ibid., p. 18. This is what Russell called a 'denoting function'.
158 *F.u.B.*, p. 18.
   Cf. also Frege's reply ('Ueber Begriff und Gegenstand', *Vierteljahrsschr. für wiss. Philos.*, xvi (1892), pp. 192–205) to Benno Kerry's criticisms. For Russell's criticisms on Frege and Kerry, see *R.Pr.*, pp. 505–10, 520–2.
159 *F.u.B.*, p. 19.

thinking that we are dealing with objects, appears to be a commandment of scientific rigour.' In the requirement that a combination of signs shall never become devoid of signification, whatever objects some or all of those signs may represent, 'we have,[160] for concepts, the requirement that they have a truth-value for each argument, that for every object it is determined whether it falls under the concept or not; in other words: they must be sharply bounded, otherwise it would be impossible to set up logical laws about them. For every argument x, for which "$x + 1$" were devoid of signification, the function $x + 1 = 10$ would have no value, and therefore no truth-value, so that the concept "what, when increased by 1, gives 10" would not have a sharp boundary. The requirement of the sharp bounding of concepts necessitates, therefore, that functions in general have a value for every argument.'[161]

Frege then introduced what, in the *Begriffsschrift*, he called the 'line of content', but here simply 'horizontal line', as a function whose value is the true if the argument (represented by the signs following the line) is the true, but the false if the argument is the false or no truth-value at all;[162] and then the vertical line of assertion.[163] That function represented by the short vertical line of denial combined with the horizontal line was introduced,[164] and the representation of generality and the combination of hollowing with a line of denial at one or both sides of it, the whole bearing or not the sign of assertion,[165] were explained in a manner essentially the same as in the *Begriffsschrift*. The form $f(\alpha)$ preceded by a hollowed out horizontal line, in the hollow of which an α stands, and on both sides of which hollow are lines of denial, occurs in the expression of existence-theorems, and Frege[166] regarded it as the expression of a function whose argument is indicated by 'f'. 'Such a function is obviously fundamentally different from those considered hitherto; for as its argument only a function can occur.' Frege called the new functions functions of the *second stage (Stufe)*, and, correspondingly, distinguished concepts of the *first* and *second degree*.[167]

---

160 *F.u.B.*, p. 20; cf. *Gg.*, i, p. 2; and ii, pp. 69–72.
161 Cf. article 12 above and the section on Peano below. Frege did not say how $a + b$ (for example) is to be defined without a hypothesis that a and b are numbers, and Russell appears to have been the first to give such a definition, '$a + b$' denoting the null-class except when a and b are numbers.
162 *F.u.B.*, p. 21; *Gg.*, i, p. 9. Thus '$1 + 3 = 4$', '$1 + 3 = 5$', and '4', each preceded by a horizontal line, denote the true, the false, and the false respectively.
163 *F.u.B.*, pp. 21–2; *Gg.*, i, p. 9. The line of judgement cannot be used for the formation of a functional expression, because it does not serve, in combination with other signs, for pointing out an object. An assertion indicates (*bezeichnet*) nothing, but asserts something.
164 *F.u.B.*, pp. 22–3.
165 ibid., pp. 23–6.
166 ibid., pp. 26–7.
167 Cf. *Gl.*, p. 65, where *Ordnung* replaced *Stufe*. The ontological proof of the existence of God treats existence as if it were a concept of the first degree.

In analysis, definite integrals are functions of the second degree (functions of functions).

"The sign '$\displaystyle\int_0^1 \frac{d\alpha}{1 + \alpha^2}$' denotes a number, '$\displaystyle\int_0^t \frac{d\alpha}{1 + \alpha}$' denotes a function of the first

Coming to functions of two arguments, Frege[168] called those functions whose value is always a truth-value, relations (*Beziehungen*); considered[169] the function of two truth-values which we know as implication (the false if the true is the *y*-argument and an object which is not true is the *x*-argument; and otherwise the true); and distinguished degrees within functions of two arguments.[170]

— 20 —

We may sum up the advances made by Frege from 1879 to 1893 as follows:

Firstly, the hypothesis as to the judicability of contents was dropped. We have seen, for example, how, in *Function und Begriff*, the relation of implication between *p* and *q* may hold when they are not propositions at all. And we there had the evils of definition under an hypothesis stated.

Secondly, Frege recognized (first in the *Grundlagen*) that equality in mathematics is always expressible by a logical identity; and hence that the sign '$\equiv$' of the *Begriffsschrift* could be replaced by the usual '$=$'.

Thirdly, the traces of formalism in the *Begriffsschrift* vanished: a function ceased to be called a *name* or *expression*.

Fourthly, the sign '$=$' ceased to be considered as the expression of a relation between *names*, and the distinction between the meaning and denotation of a name had further consequences in such fundamental matters as the analysis of judgement.

---

stage, if the letter 'ζ' holds the argument-place open, and '$\int_0^1 \varphi(\alpha)d\alpha$' denotes a function of the second stage if the letter 'φ' holds the argument-place open.

$$\int_0^1 \frac{d\alpha}{1 + \alpha^2}$$

is the value of this function of the second stage for the function

$$\frac{1}{1 + \zeta^2}$$

as argument. The expression 'function of a function' is already used in another sense. But properly speaking, this usage is false. Out of the functions $\sqrt{\zeta}$ and $1 + \zeta^2$ we can form the function $\sqrt{(1 + \zeta^2)}$. Here the value of the function $1 + \zeta^2$ enters for an argument which, for its part, is an argument of the function $\sqrt{\zeta}$. But this value is a number and not a function. Thus the argument of the function $\sqrt{\zeta}$ is by no means a function but a number. Since the function $\sqrt{\zeta}$ is of the first stage, it can never have a function as an argument. Many mathematical writers use the word 'function' where what they mean is the value of the function for a certain argument. Hence the false manner of using the expression 'function of a function'. In $\sqrt{\varphi(\zeta)}$ we have a function of the second stage, if the letter 'φ' holds open the argument-place. Let us take, in succession, as arguments, the functions $1 + \zeta^2, 1 - \zeta^2, 1/\zeta$; then we get as values of the function of the second stage the values $\sqrt{(1 + 1^2)}, \sqrt{(1 - 1^2)}, \sqrt{(1/1)}$ respectively. But it is clear that this function of the second stage is quite different from the function of the first stage $\sqrt{\zeta}$." (F., 1910.)

168 *F.u. B.*, p. 28.
169 ibid., p. 28.
170 ibid., pp. 28–30.

And the three most conspicuous advances made in the *Grundgesetze*, of which
we are about to speak, seem to have been the emphasis on the distinction between
all and *any*, the full import of which hardly appears in Frege's earlier writings on the
subject of generality; the introduction of a sign for expressing the definite article;
and the discussion of the doctrine of definition.

— 21 —

Frege's logical and arithmetical doctrines were the most fully exposed in his
Grundgesetze der Artihmetik, the first part of which was published in 1893.[171] It
was his intention[172] to establish the view on the concept of enumeral which he had
stated in the *Grundlagen*, and which rested on the result[173] that the number-datum
contains a statement about a concept.

The fundamental signs used in the *Begriffsschrift* occur again in the
Grundgesetze, with one exception: instead of the three parallel lines, Frege chose the
usual sign of equality, since he had convinced himself that the words 'equal to' in
arithmetic have the same denotation as the words 'identical with'[174]. To the
fundamental signs of the *Begriffsschrift* were here added a smooth breathing,
referred to above, to denote the range of a function,[175] and a sign to represent the
definite articles of language.[176] Development of Frege's logical views necessitated
different explanations of the original signs, whose form and algorithm remained
unaltered or nearly so.[177]

The proofs[178] contained in the *Grundgesetze* contained no words, but were
carried through with Frege's signs alone, and the progress from one theorem to the
next was according to definite rules.[179] The theorems – which there must be – which
cannot be deduced from others are the fundamental laws,[180] and, besides these,
definitions[181] also form part of the irreducible residuum. The need of definitions
appears constantly as such an investigation as this progresses; they are not creative
but merely nominal, and introduce shorter names with which we can theoretically
dispense.[182] Frege's aim in this work was, firstly, to denote all the theorems which
are used without their being proved, expressly as such, in order to see clearly on
what the whole structure rests; secondly, to reduce the number of these primitive
laws as much as possible, by proving all that is provable;[183] thirdly, to enumerate

---

171 *Grundgesetze der Arithmetik. Begriffsschriftlich abgeleitet von Dr G. Frege*,
Bd. i (Jena 1893), xxxii, 254 pages; Bd. ii (Jena 1903), xvi, 266 pages.
   Frege (*Gg.*, i, p. xi) referred to the neglect of his work by mathematicians – a
state of things which is fortunately changed at the present time.
172 *Gg.*, i, pp. viii–ix; cf. p. 1. The *Grundgesetze* had as purpose to prove the thesis
made probable in the *Grundlagen*, that arithmetic is merely a branch of logic.
173 *Gl.*, p. 59.
174 *Gg.*, i, p. ix; ii, p. 71n. Cf. *Gl.*, pp. 73, 74, 76; *S.u.B.*, p. 25.
175 *Gg.*, i, pp. ix–x, 14.
176 ibid., pp. ix, 19.
177 ibid., p. x.
178 ibid., pp. v–vi.
179 Collected in *Gg.*, i, pp. 61–4.
180 Collected in *Gg.*, i, p. 61 (they are six in number).
181 Collected in a table on pp. 240–1 of *Gg.*, i.
182 On definitions and the rules for definition, see *Gg.*, i, pp. vi, xiii–xiv, 51–2; ii,
pp. 69–80. Cf. above.
183 Cf. Dedekind, *Was sind und was sollen die Zahlen?* [Brunswick 1888].

in advance all the manners of reduction used.[184] In this way, for the first time, the necessary material for judging that arithmetic is only a further developed logic was brought together.[185]

— 22 —

One great merit of the ideography of Frege was that every sign occurring in it was explicitly defined or explained. Thus the ideas made use of in all languages, which are represented by commas and other marks of punctuation, were implicitly presupposed in practically all mathematical literature, even in so fundamental a work as Dedekind's *Was sind und was sollen die Zahlen?*; and Frege[186] called attention to this. Shortness may be attained in an ideography at the expense of completeness, but such shortness was not aimed at by Frege.

Frege's work thus differed, by its thoroughness, from the work of all his predecessors. He said:[187] 'If I compare arithmetic to a tree which unfolds at the top into a multitude of methods and theorems while the roots go deep into the soil, then the drive towards depth seems to me weak, at least in Germany. Even in a work one would like to include in this movement, like the *Algebra of Logic* by Mr E. Schröder, the drive to the top soon gains the upper hand, even before a very great depth has been reached, and effects an upward twist and an unfolding into methods and theorems.'

Frege[188] criticized the notion which mathematicians denote by the word 'aggregate' (*Menge*), and particularly the views of Dedekind and Schröder. Neither of these authors distinguished the subordination of a concept under a concept from the falling of an object under a concept;[189] a distinction upon which Peano rightly laid so much stress, and which is, indeed, one of the most characteristic features of Peano's system of ideography.

— 23 —

Frege's undertaking was, to a certain extent, analogous to Dedekind's, but was carried out with far greater accuracy and profundity. Thus, while Dedekind, at least in intention, made 'System' and 'Imaging' the two foundation-stones of his theory of arithmetic, those of Frege's theory were[190] the more precisely expressed 'Concept' (*Begriff*) and 'Relation' (*Beziehung*). In his explanation of what is meant by the words 'Function', 'Concept', and 'Relation';[191] Frege's point of view was that of his *Function und Begriff* of 1891 and his *Begriff und Gegenstand* of 1892, and was different from that of the *Begriffsschrift*.

The analysis of Frege's *Grundgesetze* will be continued later.

---

184 *Gg.*, i, p. vi.
185 *Gg.*, i, p. vii; cf. *Gl.*, pp. 99, 102–03; *F.u.B.*, p. 15.
186 *Gg.*, i, pp. vii–viii.
187 ibid., p. xiii.
188 ibid., pp. 1–3.
189 ibid., p. 2; cf. Frege's article 'Kritische Beleuchtung einiger Punkte in E. Schröder's *Vorlesungen über die Algebra der Logik*', *Archiv für systematische Philosophie*, i (1895), pp. 433–56; *B.P.*, p. 371; and the section on Peano below. Peano introduced his relation ε in 1889, and seems to have anticipated Frege. The note in Russell's *Principles*, p. 19, stating that Frege distinguished, in his *Bs.*, between what Peano denoted by ε and 'ɔ', appears to be a mistake.
190 *Gg.*, i, p. 3.
191 ibid., pp. 5–8.

CHRONOLOGICAL TABLE OF CORRESPONDENCE
CONTAINED IN THIS VOLUME

1902	7.9	Jourdain to Frege	VIII/1
1902	23.9	Frege to Jourdain	VIII/2
1902	23.9	Frege to Russell	XV/10
1902	29.9	Russell to Frege	XV/11
1902	20.10	Frege to Russell	XV/12
1902	12.12	Russell to Frege	XV/13
1902	28.12	Frege to Russell	XV/14
1903	7.1	Peano to Frege	XIV/12
1903	18.1	Pasch to Frege	XIII/4
1903	20.2	Russell to Frege	XV/15
1903	21.5	Frege to Russell	XV/16
1903	24.5	Russell to Frege	XV/17
1903	21.6	Korselt to Frege	IX/1
1903	27.6	Korselt to Frege	IX/2
1903	30.6	Korselt to Frege	IX/3
1903	7.11	Hilbert to Frege	IV/9
1904	11.2	Couturat to Frege	II/6
1904	17.3	Vailati to Frege	XVII/1
1904	21.3	Frege to Jourdain	VIII/3
1904	22.3	Jourdain to Frege	VIII/4
1904	13.11	Frege to Russell	XV/18
1904	12.12	Russell to Frege	XV/19
1905	7.1	Pasch to Frege	XIII/5
1906	21.10	Couturat to Frege	II/7
1906	30.10–1.11	Frege to Husserl	VII/3
1906	13.11	Pasch to Frege	XIII/6
1906	9.12	Frege to Husserl	VII/4
1909	28.1	Jourdain to Frege	VIII/5
1909	15.2	Jourdain to Frege	VIII/6
1910	16.4	Jourdain to Frege	VIII/7
1910	23.4	Jourdain to Frege	VIII/8
1912	9.6	Frege to Russell	XV/20
1913	29.3	Jourdain to Frege	VIII/10
1914	15.1	Jourdain to Frege	VIII/11
1914	28.1	Frege to Jourdain	VIII/13
1917	31.1	Frege to Dingler	III/1
1917	2.2	Dingler to Frege	III/2
1917	6.2	Frege to Dingler	III/3
1917	26.2	Dingler to Frege	III/4
1917	27.6	Dingler to Frege	III/5
1917	4.7	Frege to Dingler	III/6
1917	10.7	Dingler to Frege	III/7
1917	21.7	Frege to Dingler	III/8
1917	1.8	Frege to Dingler	III/9
1918	17.11	Frege to Dingler	III/10
1919	24.8	Frege to Linke	XI/1
1925	24.4	Hönigswald to Frege	V/1
1925	26.4–4.5	Frege to Hönigswald	V/2

Undated Letters

after 1891	Frege to Peano	XIV/1
after 1896 (printed 1899)	Peano to Frege	XIV/8
after 1896	Frege to Peano	XIV/11
probably 1902	Frege to Huntington	VI/1
after 23.4.1902	Frege to Jourdain	VIII/9
between 15. and 28.1.1914	Frege to Jourdain	VIII/12
after 1918	Frege to Zsigmondy	XVIII/1

The table above is extracted from a table prepared, and generously provided, by Professor Günther Patzig of Göttingen.

WORKS OF FREGE CITED IN THIS VOLUME

1 *Begriffsschrift, eine der arithmetischen nachgebildete Formelsprache des reinen Denkens* (Halle a.S. 1879). [E.T. in T. W. Bynum (ed.), *Conceptual notation and related articles* (Oxford 1972).]

2 'Anwendungen der Begriffsschrift', *Jenaische Zeitschrift für Naturwissenschaft* [hereafter JZN] XIII (1879) Supplement II, pp. 29–33. [E.T. ('Applications of the "Conceptual Notation"') as for 1.]

3 'Über den Zweck der Begriffsschrift', JZN XVI (1883) Supplement pp. 1–10. [E.T. ('On the Aim of the "Conceptual Notation"') as for 1.]

4 *Die Grundlagen der Arithmetik* (Breslau 1884). [E.T. by J. L. Austin, *The Foundations of Arithmetic* (Oxford 1950).]

5 [review of] H. Cohen, *Das Princip der Infinitesimal-Methode, und seine Geschichte* (Berlin 1883) in *Zeitschrift für Philosophie und philosophische Kritik* [hereafter ZPPK] LXXXVII (1885) pp. 324–9.

6 'Über formale Theorien der Arithmetik', JZN XIX (1886) Supplement pp. 94–104. [E.T. ('On Formal Theories of Arithmetic') as for 18.]

7 *Funktion und Begriff* (Jena 1891). [E.T. ('Function and Concept') in P. Geach and M. Black (tr. and ed.), *Translations from the philosophical writings of G. Frege* second edition (Oxford 1960).]

8 'Über das Trägheitsgesetz', ZPPK MXVIII (1891) pp. 145–61. [E.T. ('On the Law of Inertia') by H. Jackson and E. Levy in *Studies in History and Philosophy of Science* II (1971–2) pp. 195–212.]

9 Über Sinn und Bedeutung', ZPPK C (1892) pp. 25–50. [E.T. ('On Sense and Reference) as for 7.]

10 'Über Begriff und Gegenstand', *Vierteljahrsschrift für Wissenschaftliche Philosophie* XVI (1892) pp. 192–205. [E.T. ('On Concept and Object') as for 7.]

11 *Grundgesetze der Arithmetik*, vol. I (Jena 1893). [Partial E.T. by M. Furth, *The Basic Laws of Arithmetic* (Berkeley and Los Angeles 1964).]

12 'Le nombre entier', *Revue de Métaphysique et de la Morale* III (1895) pp. 73–8. [E.T. ('The Whole Number') in *Mind* LXXIX (1970) pp. 481–6.]

13 'Kritische Beleuchtung einiger Punkte in E. Schröder's Vorlesungen über die Algebra der Logik', *Archiv für systematische Philosophie* I (1895) pp. 433–56. [E.T. ('A Critical Elucidation of some Points in E. Schroeder's *Algebra der Logik*') as for 7.]

14 'Lettera del Sig. G. Frege all'Editore', *Rivista di Matematica* (*Revue de Mathématiques*) VI (1896–9) pp. 53–9.

15 'Über die Begriffsschrift des Herrn Peano und meine eigene', *Berichte über die Verhandlungen der Königlich Sächsischen Gesellschaft der Wissenschaften zu Leipzig*, Mathematische-Physische Klasse XLVIII (1897) pp. 361–78. [E.T. by V. H. Dudman ('On Herr Peano's Begriffsschrift and My Own') in *Australian Journal of Philosophy* XLVII (1969) pp. 1–14.]

16 *Über die Zahlen des Herrn H. Schubert* (Jena 1899).

17 *Grundgesetze der Arithmetik*, vol. II (Jena 1903). [Partial E.T. as for 11.]

18 'Über die Grundlagen der Geometrie', *Jahresbericht der Deutschen Mathematiker-Vereinigung* [henceforth JDMV] XII (1903) pp. 319–24. [E.T. by E.-H. Kluge, *On the Foundations of Geometry and Formal Theories of Arithmetic* (London and New Haven 1971).]

19 'Über die Grundlagen der Geometrie II', JDMV XII (1903) pp. 368–75. [E.T. as for 18.]

20 'Was ist eine Funktion?' *Festschrift Ludwig Boltzmann gewidmet zum 60.Geburtstage* (Leipzig 1904), pp. 656–66. [E.T. ('What is a Function?') as for 7.]

21 'Über die Grundlagen der Geometrie I', JDMV XV (1906) pp. 293–309. [E.T. as for 18.]

22 'Über die Grundlagen der Geometrie (Fortsetzung) II', JDMV XV (1906) pp. 377–403. [E.T. as for 18.]

23 'Über die Grundlagen der Geometrie (Schluss) III', JDMV XV (1906) pp. 423–30. [E.T. as for 18.]

24 'Antwort auf die Ferienplauderei des Herrn Thomae', JDMV XV (1906) pp. 586–90. [E.T. ('Reply to Mr Thomae's Holiday Chat') as for 18.]

25 'Die Unmöglichkeit der Thomaeschen formalen Arithmetik aufs Neue Nachgewiesen', JDMV XVII (1908) pp. 52–5. [E.T. ('Renewed Proof of the Impossibility of Thomae's Formal Arithmetic') as for 18.]

26 'Schlussbemerkung' JDMV XVII (1908) p. 56. [E.T. ('Concluding Remarks') as for 18.]

27 'Der Gedanke. Eine logische Untersuchung', *Beiträge zur Philosophie des deutschen Idealismus* [henceforth BPhdI] I (1918) pp. 58–77. [E.T. ('Thoughts') in P. T. Geach (ed.), *Logical Investigations* (Oxford 1977).]

28 'Die Verneinung. Eine logische Untersuchung', BPhdI I (1918) pp. 143–57. [E.T. ('Negation') as for 27.]

29. 'Logische Untersuchungen. Dritter Teil: Gedankengefüge', BPhdI III (1923) pp. 36–51. [E.T. ('Compound Thoughts') as for 27.]

30 *Begriffsschrift und andere Aufsätze*, Second edition, ed. I. Angelelli (Darmstadt and Hildesheim 1964).

31 *Kleine Schriften*, ed. I. Angelelli (Darmstadt and Hildesheim 1967).

32 *Nachgelassene Schriften*, ed. H. Hermes, F. Kambartel and F. Kaulbach (Hamburg 1969). [All references are to E.T., *Posthumous Writings*, tr. Peter Long and Roger White (Oxford 1979).]

INDEX

The Introduction is ignored in the following and references to footnotes (other than Jourdain's in the Appendix) are indicated by sloped figures. ~ signifies repetition of the catchword or of that part of it preceding /.